TOP
TEN
IDEAS
OF
PHYSICS

TOP
TEN
IDEAS
OF
PHYSICS

FOUNDATIONS FOR
UNDERSTANDING
THE UNIVERSE

A. ZEE

Princeton University Press

Princeton and Oxford

Requests for permission to reproduce material from this work
should be sent to permissions@press.princeton.edu

Published by Princeton University Press
41 William Street, Princeton, New Jersey 08540
99 Banbury Road, Oxford OX2 6JX

GPSR Authorized Representative: Easy Access System Europe - Mustamäe tee 50, 10621 Tallinn, Estonia, gpsr.requests@easproject.com

press.princeton.edu

All Rights Reserved

ISBN 978-0-691-22580-7
ISBN (e-book) 978-0-691-22581-4

British Library Cataloging-in-Publication Data is available

Editorial: Ingrid Gnerlich and Whitney Rauenhorst
Production Editorial: Karen Carter
Text and Jacket/Cover Design: Wanda España
Production: Jacqueline Poirier
Publicity: Matthew Taylor
Copyeditor: Jennifer Harris

Jacket/Cover Credit: Jacket image by Eric De Giuli, physicist and artist

This book has been composed in Avenir and Times

Printed in the United States of America

10 9 8 7 6 5 4 3 2 1

To Isaac Newton, James Clerk Maxwell, and Albert Einstein

TABLE OF CONTENTS

PROLOGUE

The everlasting and the coldest

I urge you to read this prologue first to get a sense of what this book is about. I do not want to disappoint those readers looking for the latest and the hottest; they would be better off perusing the web.

I have been mulling over this book for at least thirty, perhaps forty years, not long after I wrote my first professional book, *Unity of Forces in the Universe*, and my first popular book, *Fearful Symmetry*.[1] In writing these books I thought about how unity and symmetry furnish two of the foundational ideas in theoretical physics. Over the years I have sat down and written out a list of what I consider to be the top ten ideas of theoretical physics. I was also partly inspired by a late night American television program in the 1980s during which the host would promulgate a top ten list, often consisting of ludicrous absurdities. I would make up my own list and post it outside my office, sometimes just to tease my colleagues, such as the ten ugliest terms in condensed matter physics.

More recently, while composing a proposal to Princeton University Press and while writing this book, I found five or six of these lists of the top ten ideas of theoretical physics scribbled down over the decades. Interestingly and reassuringly, and to my delight as well, the lists were almost identical, with some minor changes and with some shifts in the order in which these ideas appear. Fads and fashions in theoretical physics come and go, but the foundational ideas abide. I was much gratified, knowing that these ideas on my list have lasting values, at least for as long as they had been rattling around in my mind.

In short, the focus of this book is on the intellectual foundation of theoretical physics, not ephemera. Even highly educated intellectuals with an active interest in physics are often swept away by the gee whiz discoveries, some of which merely provoke yawns among physicists and some are mercifully soon forgotten. The popular media makes the situation worse. The financial incentive is clearly to hype wild speculations that only a tiny minority of physicists actually believe in or care about. A glance at this book's table of contents would indicate that you won't find the latest and the hottest in this book. Only the everlasting and the coolest.

Ten foundational ideas of physics

In this book, I will discuss ten foundational ideas, each explained briefly in a separate chapter. Let me summarize each of these ten chapters in one or two sentences.

1. The physical world is comprehensible.
   That the universe is comprehensible is the most incomprehensible idea the human race has ever entertained. Nature has shown her kindness to theoretical physicists again and again.

2. The laws of physics are the same here, there, and everywhere, the same yesterday, today, and tomorrow.
   Physics is universal and eternal, without which we could not understand the universe, including the time when it was but an infant.

3. The world is quantum.
   Surely the most stunning discovery ever by humans is that beneath our everyday world lies a mysterious quantum world, in which common sense and the traditional notion of reality do not hold.

4. Quantum fields forever: Einstein's total love.
   The universe is pervaded by an intertwining dance of the quantum fields. Excitations in these eternal fields generate the particles we know and love, some ephemeral, others apparently eternal but perhaps not.

5. Fearful symmetry: a universe full of symmetries.
   Why should physics become more symmetric, and hence simpler and "more elegant" the more layers we peel back? A truly profound mystery.

6. Einstein, the exterminator of relativity and the choreographer of space-time.
   The laws of physics must not depend on the physicist: truth is not relative, contrary to what some philosophers would want you to believe.

7. Unity of forces in the universe.
   Physics at one point consists of myriads of laws. An almost incredible achievement of 20th century physics is to distill these laws into four fundamental interactions, and then to unify three out of the four into

a single grand unified theory, with gravity still refusing to join in the dance.

8. The Creator speaks the language of mathematics.
Wigner wondered out loud whether or not the effectiveness of mathematics in physics was unreasonable. I, and many, but not all, theoretical physicists stand perplexed.

9. Entropy and thermal agitation: all about sharing.
The interaction between two particles knows not the collective behavior of zillions of particles: entropy and the arrow of time emerge. "More is different."

10. Physics is where the action is.
Fermat's least time principle launched a movement away from the Newtonian conception of a bunch of equations of motion dictating what will happen from one instant to the next. Physics at its deepest is formulated in terms of an action with which Nature choreographs the universe.

So, ten chapters, each devoted to a big idea. Incidentally, at one point I also considered a tongue-in-cheek chapter 11, titled "Physics in Crisis: Cautionary Tales of Moral Bankruptcy."[2] That chapter has disappeared, but who knows, it may pop up in a future book.

Ten is of course an evolutionary accident on our particular planet. I may well add or subtract in the future.

Not just plug and chug, or this and that

Physics is not just about plugging in equations and chugging away to obtain a result. But after years of homework exercises in high school and then in undergraduate and graduate courses, it is certainly natural for students to get this impression. The quest for results motivates many physicists, but definitely not all. Dick Feynman perhaps put it best, in his characteristic fashion: "Physics is like sex: sure, it may give some practical results, but that's not why we do it." Results may pave the road to fame and fortune. But many of us are driven instead by a thirst to understand the inner working of the universe. Some of my friends and I are fascinated by the logical interplay between ideas that, astonishingly enough, enable our puny selves to manage a grasp of this universe, stretching from the very small to the very large. A universe vast and intricate almost beyond imagination! The phrase "It boggles my mind!" appears to me to be coined for physics.

Einstein said, "I want to know how God created this world. I am not interested in this or that phenomenon. I want to know His thoughts, the rest are details." Well said! But perhaps only Einstein has the status to dismiss most of physics as mere details. I loved this quote when I first saw it, and I used it in my first popular book, *Fearful Symmetry*. I was quite taken aback by some of the mail I received challenging me to mortal combat. There was one guy who had devoted his entire life to this, and of course he was angry as hell. And then there was another guy who knew everything about that, a world authority of that, but who had never even heard of this. Well, Einstein had been dead for decades and unavailable to pay for his hubris. But, perhaps with a twinkle in his eyes, he might have suggested that I put Doctor This and Professor That in touch with each other.

Relax. All physicists work on "this or that phenomenon," including the grand old man himself. While his most celebrated papers addressed fundamental issues of space and time and the nature of the quantum world, I also read with delight his insights on the viscosity of wet sand[3] and on the thermodynamics behind a new kind of refrigerator. As a matter of fact, almost all physicists[4] are working almost all the time on what Einstein dismissed as this or that.

Indeed, almost all the great discoveries were brought about by studying this and that. The young Max Planck ushered in the quantum revolution, emphatically not by mooning philosophically over whether or not there might be a layer of reality under the reality we know, but by studying the electromagnetic waves trapped inside heated cavities. For another example, the deduction that dark energy drives cosmic expansion depends on a detailed understanding of supernova explosions. As I have said elsewhere, physics is a big tent accommodating all kinds, from lion tamers to trapeze artists to acrobatic clowns.[5] Physics progresses on many fronts; fundamental insights about the nature of space and time by necessity could strike but once in a blue moon, contrary to what many fascinated by physics think. Yes, I know that "more is different," better than you might think.

Nevertheless, the book you are holding in your hands is about fundamental ideas in physics, and not particularly concerned with Einstein's "this or that" except as needed for illustrative purposes. So, remember this before you send off that indignant email. Of course I know that the devil is in the weeds and the importance of learning how to get down in the weeds, but this book is all about a sweeping overview.

Since this book touches on some of the topics treated in my previous books (listed in a brief bibliography given at the end of this book), I often refer to them, using abbreviations given in the bibliography.

Disclaimers

First, a few disclaimers. This is neither a descriptive popular book nor a textbook. An even minimally detailed exposition of each of these ten ideas would merit a book. For instance, I have written a popular book, a textbook, and a semipopular book on quantum field theory, the subject of chapter 4. I often have to seek a balance between brevity and totally confusing the reader. Many essential topics are omitted or simply glossed over.[6] I am brave enough to suffer the ire of certain colleagues.

My goal is to convey to you the importance of these ideas, and to give you a flavor of how these ideas came about. Sometimes I choose to focus on a narrow topic. For instance, given that there is no way no how for me to cover quantum physics in one chapter, I decided to focus on Bell's inequality, partly because, as I was working on this book, the Nobel prize was given to the three experimenters who verified Bell's inequality and upheld quantum physics. But to explain Bell's inequality, I have to introduce the reader to electron spin, perhaps one of the most bizarre concepts in physics. But yet, much of our technological society now depends on it. As another example, considering how much confusing talk clouds the popular understanding of entropy and information, I gave a detailed, almost textbook style, exposition of these two intertwining concepts.

I mentioned that I have published a number of popular books and textbooks on physics, and so, necessarily, I often find myself repeating what I have said elsewhere. Why repeat all this? Well, Feynman had asked himself this same question long before me, and I liked his answer when I was a kid. "Because there are new generations born every day. Because there are great ideas developed in the history of man, and these ideas do not last unless they are passed purposely and clearly from generation to generation."[7]

No doubt, various experts would find much in these sweeping surveys to incur their displeasure. But if I were to put in all the necessary caveats, this would not be a book, but a library. The typical reader would also be overwhelmed.

I am also more than sure that many physicists, if asked to list the top ten ideas in physics, would produce lists that differ from mine. All I can say is that my list reflects my own journey in theoretical physics, and a book of this kind is necessarily, and preferably, unapologetically opinionated. If you vehemently disagree with my list, I cordially invite you to write your own book.

This book does not present a linear narrative. I have my own reason for ordering these ten ideas, and as I said, the order has evolved somewhat. So, if you do not understand something, just keep going. Nothing requires you to master one chapter before going on to the next. These foundational ideas, however, forcibly intertwine. For instance, that the laws are revealed to be more symmetric the deeper we go (chapter 5) and the ultimate unity of apparently distinct forces (chapter 7) are perforce related. Some readers familiar with my other books would know that I favor a profusion of endnotes.* Some of these may lead the reader to delve deeper, others might merely provoke a chuckle. You may choose to ignore the endnotes upon a first reading, and come back to them later. A matter of preference.

Finally, I am not a historian, and no claim (of course) is made regarding the historical veracity of the narrative.

For whom does this book toll

In several of my popular books, I include a section in the preface titled "For whom is this book for." I would of course be immensely pleased if some high school or college students would be stimulated by the ideas outlined in this book to go on to devote their lives to physics. I also know that, at the other end of spectrum, a multitude of professionals, some retired and some not, have the habit of reading popular books about physics. This book is also dedicated to them and to their fighting spirit and urgent desire to understand the universe at a deeper level.

Clearly, some prior exposure to Einstein's ideas about spacetime and to quantum physics would be preferable; I could give only the most cursory overview of these revolutionary developments now well over a century old. In short, this book is for those intellects with a capital I, young or old, who want to get a flavor of what physics is all about.

---

*Readers, particularly the young, are urged to look through some of the endnotes, where some more mathematical discussions are hidden. There is also career advice should you go into physics. See, for example, the note about what Heisenberg said about Kronig in chapter 3.

Notes

[1] The book was first published by Macmillan and then later republished by Princeton University Press in the Princeton Science Library series. I am proud to say that it has since been translated into German, Japanese, Korean, French, Spanish, Turkish, and Chinese (in both traditional and simplified characters). I am extremely gratified over the years, particularly during my travels, to have met a stream of physicists who have read this book early in their intellectual journey.

[2] For those readers not residing in the United States, chapter 11 of the bankruptcy code is commonly equated with bankruptcy.

[3] Once long ago, an oil exploration company even paid me to read this rather obscure contribution by Einstein. Thanks, Joel.

[4] And those who haven't are known, at least to me, as mathematicians or philosophers.

[5] Sadly, in our degenerate era, the circus is often dominated by clowns.

[6] This is partly to forestall those on Amazon who might confuse textbooks and popular books.

[7] R. P. Feynman, *The Meaning of It All*, 1963.

TOP
TEN
IDEAS
OF
PHYSICS

1
THE PHYSICAL WORLD IS COMPREHENSIBLE

> The most incomprehensible thing about the world is that it is comprehensible.[1]

Why is the world comprehensible?

Of Einstein's many insightful sayings, this is my favorite. Indeed, this apparent miracle, that the world is comprehensible, is what motivates physicists. In a lawless universe behaving with neither rhyme nor reason, physics would be impossible.

I should hasten to say what hardly needs to be said, but with silly nitpickers lurking about I have to say it. What Einstein meant by the world is the physical world, the world studied by physicists. Plenty of questions are beyond the comprehension and grasp of physics, of course. What is the meaning of life? Why are we here at all? Why does the universe exist? Is there another level of existence we know nothing about? You could make up your own list.

We have made stupendous progress since our days in the caves when nearly everything was incomprehensible. But even within almost living memory, the progress has been so vast that it almost beggars imagination. Two hundred years ago, physicists did not know what light was. One hundred and fifty years ago, physicists had no inkling that spacetime was curved. One hundred years ago, physicists struggled to grasp the laws governing the realm of the quantum underlying our world. It is true that progress in theoretical physics has slowed nowadays. Nevertheless, progress is being made.

Why should Nature be comprehensible by the mind of these creatures that evolved recently on a dust mite of a planet orbiting an insignificant star drifting in a far-from-prestigious neighborhood of a mediocre average-looking galaxy? Why should the laws of physics be simple and beautiful? We could have found ourselves living in an intrinsically ugly universe, a chaotic world in no way graspable through thinking, as Einstein, he again, put it.

A smashingly revolutionary idea

Some readers might be surprised by my choice of the idea that tops my list of top ideas. The world is comprehensible! We tend to take for granted that physics is possible, that the physical world is comprehensible. But as far as I know, no ancient civilization, East or West, harbored the idea that the world was comprehensible, except, arguably, for a handful of Greeks. Even in the West, this enormous and profound idea surfaced only in the last four or five centuries, and the first inkling that the world could be understood without invoking the divine was met with ferocious resistance. It was a slow, gradual awakening threatened with excommunication, torture, and death.

The sun god pulling his chariot daily across the sky from east to west, and then somehow teleporting himself and his entire conveyance back to the east at night—this was our "understanding" not all that long ago. Yet now, for the first time in human history, a reasonably advanced student of physics could open a book on astrophysics and learn how to calculate the temperature and pressure inside a star, a particularly simple homework exercise for a garden-variety star like the sun, neither a red giant nor a white dwarf, without too many peculiar features.

Why versus how

I cannot tell you why the world is comprehensible. It is important to understand that physics does not answer the why questions. Physicists strive to steadily turn the whys into hows. We cannot tell you why the apple falls, but we can tell you how it falls, and reduce a variety of phenomena into an ever smaller set of underlying phenomena. Along the way, many whys were replaced by fewer whys, and whys by hows. Indeed, a crucial step in making physics possible is to separate the how questions from the why questions.

We can tell you why the sky is blue. It has to do with how light scatters off the air molecules, and after a heroic struggle we have reduced that to how two quantum particles, the photon and the electron, interact with each other. We can tell you that every massive object in the universe attracts every other massive object. We can tell you a lot about this force called gravity, and how it is equivalent to curved spacetime. But we cannot tell you why our universe contains such a force. Of course, were this force absent, then our universe would look quite different from the one we know.

The preceding sentence gives a capsule summary of the anthropic principle. It has been abused to answer every why question with the existence of humans. You know the jingle; various physical constants must have the value they have, since otherwise humans would not exist. A fascinating subject perhaps, but I will stay away from it in this book. Only the tried and true foundational ideas, as I said.

We do not know why there are four fundamental interactions in the universe, instead of two or six, but physics can tell us how three of these four interactions could be unified into a grand unified theory, as we will see in chapter 7.

That the physical world is comprehensible is miraculous

Physics as a subject wouldn't exist were the world not comprehensible.

Yet, if you think about it, the very comprehensibility of the world is miraculous. Sad that we have grown blasé in this enlightened age and take so much for granted. But flash yourself back a few millennia when no one even presumed to imagine the possibility of comprehending the world, let alone to attempt the task of actually comprehending the world.

You need only look at ancient cultures around the world. In school, we learned about the realm of rationality that is ancient Greece. Even in Greek myths, we could discern the urge to understand. But the notion that we could comprehend the physical world was far from universal. For example, the philosophers of ancient China envisaged man in harmony with Nature and emphasized understanding human behavior rather than the physical world. The West emphasized the control of Nature. Perhaps controlling is but one small step from understanding. To some extent, the entire Greco-Roman Judeo-Christian conception of Nature to be conquered and ravaged was missing in ancient China.[2]

Noticing regularity alone is not enough. Certainly, Asians recorded regularity with as much diligence as the Greeks. But the notion that there might be laws governing these regularities was mostly absent. That the world was comprehensible was far from obvious to every thinker in every civilization.

Nor should we assume that the development of physics as it happened on our planet is typical. For example, a civilization could develop in a binary star system, on one of the nineteen planets orbiting the two stars, which in turn waltz around each other in a highly eccentric ellipse. The regularity of the heavens, as well as the usual law of gravitation, might remain hidden from this particular civilization for a long time, perhaps longer than the natural lifetime of the civilization.

Or imagine us in a civilization on a planet much like Terra, except that it is completely covered by a single deep ocean. Electricity might be known as an epiphenomenon associated with some peculiar fish regarded as far inferior to *Pisces sapiens*, which is what we call ourselves scientifically. Magnetism is however totally unknown and light is merely an exotic phenomenon near the edge of the habitable world. Physicists would have developed an extensive understanding of water waves.[3] Some of our dead sink to a nether region where none of us are able to roam due to the crushing pressure. Others among our dead would float upward, and from observing these lightweight dead, politicians, celebrities, and such, some bright youngster eventually proposed the existence of a force named buoyancy. Later, a fishy Einstein would have the brilliant insight that buoyancy is due to a more fundamental[4] force called gravity pulling the water around these lightweight objects down.

We alone could understand the universe?

I do not doubt for a second what my computer science friends tell me, that with massive data analysis, artificial intelligence would in the foreseeable future allow us to communicate with animals. At that point, perhaps we could ask them what they understand of the physical world. Yet I also have no doubt that no other lifeforms on earth can understand nonabelian gauge theories (which we will encounter in chapter 7), let alone invent them. But then why is the physical world comprehensible to humans (at least thus far)?

Einstein was absolutely right that it is a profound mystery. It could well be that this particular primate species, after eons of evolution, is no more capable of fully understanding the physical world than one of the gophers in my backyard. (The emphasis is on "fully"!) It is certainly possible, of course, that we will eventually hit a brick wall. The pessimists in the physics community would say that we have already hit that proverbial wall.

I want to be optimistic because I need that optimism to buoy me professionally. Einstein expressed his profound wonderment more than a hundred years ago. He and his contemporaries could also feel, when confronted with the mysteries of the quantum world, that the striking progress in understanding of the physical world gained in the 19th century may come to a grinding halt. But they didn't, and pressed on. And progress has been made for a century and more. Of course, in deciphering the quantum world, that generation of physicists was pushed along by an almost excessive wealth of experiments, which are now sorely lacking, no matter what the cheerleading boosters try to convince us otherwise. Needless to say, I am talking about landmark experiments that reveal some deep truths, not routine experiments measuring the decay rate of an absurdly named particle or the conductivity, super or not, of some newly minted alloy.

I am among those who believe that the universe, with its zillions of galaxies each with zillions of stars, is teeming with intelligent life. I have even thought seriously about communicating with extraterrestrial intelligence.[5] When we finally come into contact with extraterrestrial intelligent beings, what can we talk about? Not terrestrial biology, which may or may not be similar to their biology.[6] And definitely not Tang dynasty poetry, nor Shakespearean plays, which so many scholars spent, are spending, and will spend entire lifetimes on. But surely, if the extraterrestrials have mastered enough technology to communicate with us, they could discuss the classification of Lie algebras* and gauge theories with us, perhaps teaching us a thing or two. Whatever, but I could hardly doubt that they have also come to realize that the universe is comprehensible.

Could have been an impenetrable mess

A priori, the laws of physics governing our universe could have been an impenetrable morass. Or they could be evolving in a horribly complicated way. Or consider the amusing science fiction scenario that our universe

---

*From a branch of mathematics called group theory and used in advanced physics. More in chapters 4, 7, and 8.

is a school-assigned experiment carried out by a high school student in a metauniverse. Perhaps he or she or it has even started a whole bunch of universes, like ant farms, and stashed them away somewhere in the basement, out of his or her parents' way. Perhaps he (that's a lowercase H) has lost interest and forgotten about his universes, leaving some to expand, others to collapse, in utter futility and silence.

In our computer age, we are perhaps more inclined to think that we may be living inside somebody's giant computer. (Whatever that means! I for one have not the foggiest idea, in spite of various popular movies.) If so, then what an elegant program this computer is running!

To the contrary, we could certainly imagine a monster of a nasty program with ten trillion lines of code governing the motion of particles we display on a 3-dimensional screen, with the motion specified by some ghastly mess of an algorithm, but consistent with Newton's laws almost all of the time. For instance, line 4,878,923,767,863 could state that when the time in an internal clock expressed in some unit and truncated to an integer is given by a prime number, the laws of motion would be altered for a brief time equal to the eleventh smallest factor of the number one higher than that prime number. Line 4,878,923,867,863, on the other hand, could state that whenever a country whose English name starts with the letter R, such as Romania, has a national holiday the laws would be reversed for 93 nanoseconds, if we were so provincial as to refer to human coined units and alphabets. The poor physicists living inside this computer would be bitterly frustrated trying to figure out what is going on.

But no! Instead, theoretical physicists have discovered that the laws governing our universe are so simple that they could be written down on the back of a small envelope, as will be described in chapters 7 and 10, and that they are described by established (and elementary) branches of mathematics such as group theory and differential geometry. Furthermore, only the introductory portion of these subjects is needed for physics: the relevant mathematics is simple enough to be mastered by bright undergraduates, as will be discussed in chapter 8. So far, there is absolutely no evidence that physics requires any theory so complicated that it will take bright graduate students several decades, rather than a mere several years, to learn. To me, both a miracle and a mystery!

Physicists from Einstein on have been awed, and mystified, by this truly profound fact that as we examine Nature on deeper and deeper levels, She appears ever more beautiful. This beauty is not just a matter of vague feel-good talk, but a beauty that could be mathematically quantified by the notion of symmetry, as I will explain in chapter 5.

To me, the comprehensibility of the universe that so boggled Einstein's mind testifies that the universe is ordered at a deep level and not thrown together with random abandon.

Nature's kindness to physicists

During my study of physics I was repeatedly struck by Nature's kindness to physicists. It often seems that Nature is taking physicists by the hand and guiding them step by step through a predetermined curriculum.

Starting around the end of the 19th century and up till the 1920s or so, theoretical physicists understood, after much confusion and controversy, that electromagnetism is described by an abelian gauge theory. (For our purposes here, it is not important to even know what this technical term means.) During the same period, spearheaded by experimental discoveries, physicists came to realize that, in addition to gravity and electromagnetism, the world contains two other fundamental interactions, the strong and the weak interactions, which manifest themselves only in the subatomic world. In the 1970s, two of these four interactions, namely electromagnetism and the weak interaction, were unified into a single interaction, now enshrined in textbooks as the electroweak interaction, described by a nonabelian gauge theory, which contains the abelian gauge theory governing electromagnetism as an almost trivial piece. Again, it is not important to know what this gobbledygook term means, besides the fact that nonabelian gauge theories are orders of magnitude more intricate and difficult to master than abelian gauge theories, as you might have guessed.[7]

We will come back to this drive toward unification in chapter 7, but here I want to impress upon you Nature's kindness to physicists. Without first having been exposed to the easy case of electromagnetism, a practice problem as it were, theoretical physicists would have been in no position to tackle the weak interaction. So it is almost as if Nature taught physicists the baby stuff first, and then gently showed them how to move on from the abelian to the nonabelian case. Why not hit them with the nonabelian stuff first and blow them back to the cave of dark ignorance?

Allow me to repeat what I said in the prologue. Readers who have gotten this far must have realized that this book is most useful to those who have had some prior exposure to theoretical physics through popular books and the media. In contrast, those who have never heard of atoms, say, might find this book tough going. In short, this book is meant for those who appreciate an overview of theoretical physics and who are not particularly interested in what Einstein called "this and that."

As easy as possible, but not any easier

As another gesture of kindness, Nature couples the electromagnetic field to the charged fields, such as the electron field, rather feebly, as measured by the fabled fine structure coupling α with the approximate value of 0.007. This teeny number enabled a bright young post–World War II generation of theoretical physicists including Feynman and Schwinger to master quantum electrodynamics. Their perturbative approach involves successive approximation in powers of α, treating the first correction and the next correction as small and even smaller,[8] proportional respectively to $\alpha \simeq 0.007$ and $\alpha^2 \simeq 0.00005$. (Some readers might have heard of Feynman diagrams, which are merely pictorial devices quantum field theorists use to keep track of successive terms in this approximation scheme.[9]) If we were so unfortunate as to live in a universe with $\alpha \simeq 7.1$ say (just to write down an arbitrary number), subsequent terms would be explosively larger, and this glorious thrust forward of physics after World War II would have been impossible.

As an older example, the orbit of the moon around the earth and the orbits of the planets around the sun, while not perfectly circular, are fairly close to being circular. Thus, physicists could master the much simpler case of a circular orbit first, and then go on to tackle elliptical orbits.[10] Students are certainly taught this way: first the circle, and then the more complicated ellipse. The eccentricity of the ellipse may be treated as a small perturbation. Why should Nature care? Physicists could have been faced with highly elliptical orbits from the very start.

Nature's kindness pops up again and again. Understanding surface water waves gave physicists the mathematical tools to understand electromagnetic waves, and from that later to quantum waves. As a 20th century example, during their forays into the atomic and nuclear world physicists realized that they had to learn a branch of mathematics known as group theory, as will be discussed in chapter 5. Nature showed them the simplest groups first. Just to give you a flavor, let me mention a few of the groups that popped up in physics in chronological order: $SO(2)$, $SO(3)$, $U(1)$, $SU(2)$, $SU(3)$, $SU(5)$, $SO(10)$. You don't have to know anything about group theory to suspect that the integers represent a progression[11] in baby steps from quasi-triviality to graduate school level stuff. I teach this stuff[12] to undergraduates every year, and even the less endowed students are capable of progressing from $SO(2)$ to $SU(3)$, while the more talented sail on easily to $SU(5)$ and $SO(10)$. In some alternative universe, Nature could easily have been unkind and blasted physicists with $SO(10)$ the moment they set foot in the quantum world.

I feel that Nature takes great care to make physics as easy as possible, but not any easier, to paraphrase Einstein.

Intertwining ideas

The top ten ideas I talk about in this book naturally intertwine with each other to form an organic whole that is fundamental physics. That the world is comprehensible could only be made possible by the existence of immutable and universal laws, to be discussed in chapter 2. We could discover the underlying laws only if physics becomes simpler and simpler as we delve deeper and deeper. This will be discussed in chapters 5 and 7. Imagine, to the contrary, a world in which physics becomes more and more complicated, messier and messier. Many physicists, I for one, would simply give up. At least thus far, there is no sign of that disastrous scenario.

Kelvin versus Newton

> There is nothing new to be discovered in physics now. All that remains is more and more precise measurement.[13]
>
> —Lord Kelvin, in 1897

> I do not know what I may appear to the world, but to myself I seem to have been only like a boy playing on the seashore, and diverting myself in now and then finding a smoother pebble or a prettier shell than ordinary, whilst the great ocean of truth lay all undiscovered before me.
>
> —Isaac Newton, ca. 1726

Newton is partly right and partly wrong. Wrong if we take his metaphor too literally. He might not have known it, but Newtonian mechanics was a huge step toward making the universe comprehensible. Many later developments, such as fluid dynamics, amount to applying Newtonian dynamics to fluids, instead of particles and rigid bodies. But of course, he was also mostly right about the vast ocean of physics that he could not possibly know anything about, such as quantum mechanics and quantum field theory.

Kelvin, on the other hand, was entirely wrong, and has ever since been mocked as the paragon of the overly optimistic physicist. His sin consists of not asking too many questions, such as what constituted matter and how did the sun keep on burning for so long. So, let us desist in deriding an otherwise great physicist. Leave Kelvin in peace and pull up another optimist, albeit lesser known than that British Lord, speaking a generation or two later.

> It is difficult for those who did not witness it to imagine the enthusiasm, nay presumptuouness, which filled our hearts in those days. I shall never forget the terse way a friend of mine (now a very eminent figure in the world of physics) expressed his view of our future prospects: "In a couple of years," he said, "we shall have cleared up electrodynamics, another couple of years for the nuclei, and physics will be finished. We shall then turn to biology." Léon Rosenfeld (1904–1974), an early quantum physicist

Consider that Rosenfeld was twenty-one when Heisenberg (less than three years his senior) first wrote down, in 1925, quantum mechanics in the form we now know it. I do not know who Rosenfeld's friend was, but almost surely he was also a youngster in his twenties. So we could chalk this up to youthful enthusiasm.[14] At least Rosenfeld suspected that he and his friend were presumptuous.

The question I like to raise is whether our situation is closer to that described by Newton or that described by Rosenfeld's friend. The correct answer is of course nobody knows, least of all I. I am a theoretical physicist, not a fortuneteller. But still, as an exercise in futility, let me try to quantify the question. Does what we know amount to 5%, or 20%, or 80%, or 95% of all the physics there is to know? (Of course, in itself, without the kind of exhausting definitions philosophers are so fond of, this question is meaningless. But I think most working physicists know what I have in mind.) Physics may well never possess a full knowledge of turbulence, but I and many others feel that we have a reasonable understanding of the physics involved. As another example, a few decades ago, a detailed theory of how coffee stains formed was published. Does that mean the physics involved was beyond the ken of Einstein? Or even that of Kelvin? Of course not. They simply did not address the details of this particular phenomenon, and physicists never finish analyzing every possible physical phenomenon we would encounter. Rather, I am talking about broad understanding of how the universe is put together—what I call fundamental physics without bothering to define that term precisely.

It would be a fool's errand to attach some meaning to the percentages I just threw out glibly, but I will take the risk just so that the reader would have at least a rough idea of what I have in mind. Consider an insect sensitive to light crawling across a computer screen showing a movie being streamed. That insect would have not the faintest glimmer of the meaning behind the patches of changing color, let alone the electronics used to produce them. It has no way of distinguishing one patch from another, of understanding that one represents the Bad, the other the Ugly. In contrast, while I do not know how a computer works, I know about electrons, the emission of light, streaming, the wild American West and all that. I have even heard computer savvy types talking about strings of 0s and 1s. Let what that insect knows and what I know represent what I mean by 5% and 80% respectively. I believe that many of my colleagues in theoretical physics, I included, tend to believe that our understanding of the universe is more like the latter rather than the former. We know quite a lot about the universe already. I find it impossible to believe that the "patches of color" we observe "merely" represent some story enacted for the amusement of the gods.

Yes, the nattering nabobs of negativity can easily sit back and compile a long list of things physicists do not know, such as what constitutes dark matter.[15] And yes, there could also be surprises over the horizon. Indeed, dark matter was unsuspected until Fritz Zwicky[16] inferred its existence in 1933 from observing the rotations of galaxies. Since then, experimentalists have spent decades of their lives searching for the particle responsible, and thus far, they have not detected any terrestrial evidence of its existence. The search goes on, and I sure hope that one of my experimentalist friends find it soon.

I could go on and on. There are lots of questions physicists have yet to answer definitively. While there are plausible theories about dark energy, it has not been pinned down.

Ironically, it is our understanding of quantum physics that bolsters our confidence in what we have yet to detect. Perhaps ironically, thanks to quantum connectedness, we could be more confident than we would be otherwise. To be concrete, when we calculate something apparently innocuous such as the magnetic moment of the electron (which measures how the electron responds to a magnetic field), the existence of a hitherto unknown particle would contribute to the theoretical value of the magnetic moment so that we can say something about its mass for instance.

To use a phrase made famous by a former secretary of defense[17] of the United States, in physics we are confronted with the known knowns, the unknown knowns, the known unknowns, and the unknown unknowns. It is of course the flaming promise of the final category that draws generations of bright young people to a lifetime of physics.

Incomprehensible comprehensibility

> Man is only a reed, the most feeble in nature, but he is a thinking reed.
> —Blaise Pascal

To conclude, what do we make of Einstein's "incomprehensible comprehensibility"? As I have already declared, we poor physicists cannot tell you why the physical world is comprehensible. For why, we have to defer to the philosophy professors. Meanwhile, you and I could also speculate over afternoon tea and cookies such as those provided at the institute for theoretical physics where I work. Whether or not you share Einstein's incomprehension of the comprehensibility of the physical world is of course completely up to you, and depends on your outlook.

There are those who believe that out of the vastness of the universe we humans are the chosen ones. They point to the fact that, also, out of the vastness of our planet we are the only ones who comprehend. We are somehow uniquely blessed and chosen by some divine creator to grasp and to admire its handiwork.

I refuse to believe our uniqueness, given the sheer number of galaxies, each containing an almost uncountably many stars. And I know of no physicists who would seriously believe this. (But that may be due to my limited circle of acquaintances. No doubt somewhere live religious zealots who also call themselves physicists.)

One possible view is that we comprehend only what we are able to comprehend, and that represents only a fraction of what there is to comprehend.[18] No idea how big or how small that fraction is! The British physicist Arthur Eddington concocted a parable[19] that left an impression on me when I first heard it as a student. Imagine a village of fishermen, among whom one with a particularly scintillating mind formulated the fundamental law of the sea stating that all fish are longer than 1 cm, not realizing that their fishing nets all have a mesh larger than 1 cm. Is it possible that the human mind filter out of that part of physical reality that we have no way of comprehending, leaving us with what we can comprehend? Notice that I was careful to qualify reality with the word "physical." Personally, I am quite willing to concede that there may be matters beyond the grasp and reach of physics.[20]

I no longer believe Eddington's parable now that I have understood more physics. Quantum connectedness would appear to make the story less plausible. To extend the parable, perhaps a weak analogy of connectedness would be big fish devouring smaller fish. What we do understand of the four interactions hangs together so seamlessly and wondrously that it is hard to believe that there is a lot more about physical reality that we have missed. I do not mean, of course, to dismiss the possibility that there may be a deeper layer in which quantities that we now treat as mere parameters, such as the masses of the quarks, could be calculated. Indeed, string theory is a possible candidate example for this deeper layer, independently of whether or not you believe it.

Do not be confused by artificial intelligence, at least as it is now known. At present, the advantages of an inanimate computer over a human are its lightning speed and its complete ignorance of mental fatigue. But what was needed to master atomic physics was not massive amount of calculation in classical physics done blindingly fast, but deep insights and synthesis of observational data leading to a new kind of physics, namely quantum physics. Similarly, realizing that gravity is a manifestation of curved spacetime is not a matter of being able to perform a mountain of calculations using Newtonian gravity.

Eventually we could imagine artificial intelligence reaching a stage in which it could consider all possible theories (whatever that may mean in practice) and investigate the consequences and implications of each of them. When we reach that stage, then perhaps we could say that artificial intelligence has developed a measure of insight. But we appear to be far from that vision, and I am not sure that the notion of "all possible theories" could even be formulated.

In the meantime, there are still, and hopefully there will be, physicists to chew over the universe.

Notes

[1] Albert Einstein "Physics and Reality," *Franklin Institute Journal* (March 1936).

[2] Sadly, China is coming late to this highly useful idea, and now a common attitude goes "You ravaged the world, now it is our turn, so don't act holier than thou and preach to us."

[3] But not the surface water waves we commonly observe and teach to some undergraduates. See *FbN*, chapters VIII.1 and 2, in particular page 282.

[4] An early example of unification that I will talk about in chapter 7.

[5] See my SETI talk and the published papers mentioned therein. https://www.youtube.com /watch?v=MHuXlJzqKqs.

[6] See, for example, A. Kershenbaum, *The Zoologist's Guide to the Galaxy: What Animals on Earth Reveal about Aliens–and Ourselves*, and A. Weir, *Project Hail Mary*.

[7] Sadly, in physics, the name of Niels Henrik Abel (1802–1829), the brilliant Norwegian mathematical prodigy who died of tuberculosis young and poor, carries the connotation of being "simple minded."

[8] See Schwinger's tombstone shown in chapter 4.

[9] See *QFT ASAP*, chapter IV.3, and *QFT Nut*, chapter I.7.

[10] Such as that of Halley's comet. Newton in his genius appeared to have done both at the same time in his published treatise, but presumably in his actual work, he mastered the circular case first.

[11] The group $SO(2)$ describes rotations in the plane, as demonstrated all the time by hands in clocks and watches.

[12] Using my textbook *Group Nut*, which is based in turn on my course.

[13] Lord Kelvin, in 1897, speech addressing the British Association for the Advancement of Science.

[14] We could hear echoes of that sentiment in the early days of string theory when it was triumphantly called the theory of everything.

[15] The mainstream view is that dark matter consists of one or more particles. I, and I dare say most theoretical physicists, find the alternative of modifying Newtonian gravity just to accommodate the observed galactic rotation speeds ad hoc and distasteful. Part of the problem in the search for dark matter is that it is too easy to cook up a theory of the dark matter. You could simply make up a particle, which interacts only with gravity and hardly with anything else, and thus could affect the rotation of galaxies yet be virtually undetectable in any terrestrial laboratory. Such a theory is possible, but rather uninteresting. In this sense, theoretical physics could be laughably easy. Don't be impressed by someone who tells you that he or she has a theory of dark matter. Incidentally, I myself published one of the early particle theories, in which the dark matter is a particle that interacts only with the Higgs sector (never mind what that means for the moment). Hopefully, actual detection and subsequent experiments will help to sort out the theoretical morass.

[16] He of the "spherical bastard" fame in physics, a prickly legend even by the combative standards of theoretical physics.

[17] Donald Rumsfeld.

[18] For an alternative view based on the decoherent histories interpretation of quantum physics, see J. B. Hartle, arXiv:1612.01952 [gr-qc] 2016.

[19] See also *Fearful*, page 280.

[20] Here I could perhaps quote E. F. Sanders from her book *Eating the Sun*: "A sense of wonder can find you in many forms, sometimes loudly, sometimes as a whispering, sometimes even hiding inside other feelings, being in love, or unbalanced, or blue."

2
THE LAWS OF PHYSICS ARE THE SAME HERE, THERE, AND EVERYWHERE, THE SAME YESTERDAY, TODAY, AND TOMORROW

The laws of physics are the same in heaven and on earth

From day one, the entire human race had looked up at the ethereal moon floating there, all the while waxing and waning like the tides. Newton alone among the multitudes realized that the moon was falling.

Yes, falling, just like the apple!

On a moonlit night, the poet inside me trumps the physicist and I can barely believe that the same law applies to that celestial apparition as to a common rock. The full moon appears so still and so round, serenely looking down at us. You are not alone if you find it difficult to visualize the truth, that the moon is constantly falling, pulled in relentlessly by the earth in the physics nerd's version of a fatal attraction. The unceasing and inexorable pull of gravity!

For interminable ages, the savants yakked about celestial mechanics and terrestrial mechanics, but after young Isaac had his blinding insight under the apple tree, there was only one mechanics: Newtonian mechanics. The falling moon and the falling apple are governed by the same laws of physics!

Most people nowadays, without thinking much about it, would regard this astonishing fact as self evident. But time travel back and imagine what courage it took to go against the intellectual tyranny of the Church. Part of the battle was already won by Copernicus and Galileo, but still, I often muse that, fortunately for physics, Newton lived in England, not in Italy or Spain.

Note that I did not say "pulled down relentlessly," but pulled in relentlessly toward the center of the earth. Down is not in the vocabulary of the Creator, merely a local concept due to our myopia, not knowing how small we were compared to the earth.

By an intricate series of steps, each of which is epochal in its magnitude, meriting either the Nobel prize or the Fields medal or both were Newton alive today, young Isaac managed to change fundamentally not only how we think about physics, but how we think about the world and the heavenly bodies. Imagine inventing mechanics, gravitation, and calculus all at once! I will go through the breathtakingly beautiful chain of reasoning[1] involved.

Acceleration versus velocity, Newton versus Aristotle

Before we get to heaven, we have to talk about force. Aristotle thought that force was needed to maintain a constant velocity. (He also thought that apples fell down because they wanted to go home.[2]) But Newton said no: force is needed only to change the velocity, to produce acceleration. Yes sir, in school we learned that $F = ma$: a force F applied to an object of mass m produces an acceleration a given by F divided by m. We have been taught that magnificent truth for so long that the truth may no longer seem so mind blowing, but imagine yourself a peasant on the farm Newton's mother married into.

As you pushed a cart loaded with cabbage now getting stuck in the mud, you might be forgiven, in between actual violence to the mule and imagined violence to the young gentleman sitting under a shady tree, to yell, "Master Isaac, what are you talking about? Aristotle is so obviously right: force is proportional to velocity. Anybody can see that you need to sweat to keep something moving at constant velocity. I have even heard that is what they are teaching in that college of yours by the river Cam!"

Meanwhile, Newton's stepfather is grumbling from a distance. "That good for nothing, sitting under the apple tree and daydreaming all day, too lazy to even stand up to pick a fruit—they have to fall on his thick skull before he takes notice."

In between hatching plots[3] to bash in his stepfather's skull, the sullen teenager thought back to last winter, about how he flung small flat stones across the frozen pond, watching them slide and slide. It almost seemed that the stones would want to slide forever.

Nowadays we understand all too well this manifestation of inertia or universal sloth: things wanting to continue doing whatever they have been doing. Without the intrusion of a force, resting objects want to continue resting, and moving objects want to keep moving. A couch potato stays a couch potato. A driver slams on the brakes, and everything inside the car goes flying forward or at least tries to, if not restrained by seatbelts. You wash your hands in a public restroom and do not see any paper towels. Not wanting to wipe your hands on your clothes, you shake them back and forth. Every time your hands come to a sudden stop, the water droplets on your hands keep on trucking and fly off. Wet dogs understand this bit of basic physics especially well.

The crucial concept Newton needed was idealization. In an astonishing feat of human imagination, Newton could see what would happen on a perfectly smooth, not just an extremely smooth, icy pond. A sliding stone would keep sliding forever.

Idealization and simplification

Physics should be made as simple as possible, but not any simpler.[4]

Trumpet blast please. What a daring and brilliant thought! The idea that one can abstract away all the grime and crime, friction and dissipation, supermarket tabloids and reality TV, and imagine an ideal world!

I wrote the passage about the icy pond long ago in the heat of a Brazilian summer, and I thought that physics is also fortunate that Newton did not live in Brazil. You need to have seen skaters gliding past, seemingly forever.

So here was the young Newton head butting Aristotle, an authority so big that he failed to see, as high school physics students are now taught to see, that friction is an unnecessary complication. Newton concludes correctly that a force is necessary to produce acceleration, but not to maintain a constant velocity. Aristotle has now been banished from reputable physics departments everywhere.

That friction may be dismissed as an unnecessary complication is one of the most astonishing ideas in the history of the human race, one that makes physics possible. Imagine that, if you had spent your entire life in say the Amazon jungle!

The difficult art of declaring what is, or is not, an unnecessary complication[5] appeared to be reserved for the very best theoretical physicists. Some physicists are mired in complications, while others throw the bath water out together with the baby without realizing it. Neither extreme is recommended.

The distinguished physicist Sam Edwards behooved theoretical physicists to simplify and simplify, to ignore all inessential complications until the problem becomes trivial, and then to go back one step. Can't do that in engineering! When thinking about the moon going around the earth, Newton ignores the sun and all the other planets. He then ignores that both the earth and the moon are spinning. He even ignores their sizes and treats them as two points attracted to each other by a mysterious force. If he also ignores this force, then the problem becomes a nonproblem, just two points freely moving about. Putting the force back in, he finally has a problem he can solve.

Physics professors probably all share my experience, that beginning students tend to ask an endless stream of "what about" questions, what about this, what about that. They typically want to make physics as complicated as possible, contrary to what Einstein advised. Idealization and simplification do not come naturally, at least to some. The caricature of the successful theoretical physicist is a person who when told to solve a problem in dairy husbandry would begin by saying, "Assume a spherical cow."[6]

Moving forward at blinding speed while falling allows the moon not to hit the ground

To say that the earth is pulling on the apple by an invisible force, not needing contact to do its thing, is already a stunning leap that few were capable of. But then to have a further mental leap saying that the same[7] force is pulling on the moon through the vastness of empty space really boggles my mind. On a sunny winter's day recently, I was sitting on a bluff overlooking the Pacific, watching a pale moon perfectly still in the crystal blue sky. It sure does not look like the moon is falling, I said to myself. It does not even seem to be moving; just floating motionless over the ocean with no landmark in sight. But yet, if I had waited a while, I would have seen the moon move, as every observer has seen, on a 24-hour cycle while the sun rises and sets and on a 28-day cycle while it changes its shape. The first type of motion is apparent, due to the observer on earth rotating, the second is the actual motion of the moon as it orbits the earth. Since the moon is so far away, we hardly perceive it as actually moving like a speeding bullet.[8]

The falling apple inspired Newton to think that the moon was also falling. You the exceptionally intelligent peasant again, muttering: "Master Isaac was nuts to say that no force is needed to keep a cart moving, but now he has really lost all his marbles. We could all see that the moon is staying up there, not falling. If the moon is falling like the apple, it would have hit the ground long ago!"

How could something be falling and not hit the ground?

Newton proposed an ingenious response: as that something is falling toward the ground, the ground is trying desperately to get away.

"Now Master Isaac has really gone out of his mind. The ground is the ground, sitting here fixed for all to see. I am ankle deep in mud standing on it right now."

Ah, Newton knew something that you the peasant[9] did not know, that the earth is round, as was long known to some of the ancients.[10] This when cleverly combined with the fact that the moon is moving forward at a blinding speed liberates the moon from hitting the ground. Were the moon falling straight "down" (whatever that means) it would of course hit the ground. And were the earth flat, even a forward moving moon will eventually hit the ground.

Newton explained this by a famous thought experiment of firing a cannon from a mountain top with ever more explosive power. At some critical speed, the cannon ball would go into orbit and not hit the ground.[11] This is the principle that allows humanmade satellites and various space stations, as well as the moon, to stay up there.

We can understand the working of the heavens

After dinner [on[12] April 15, 1726] the weather being warm, we went into the garden and drink thea, under the shade of some appletrees, only he and myself. Admidst other discourse, he told me, he was just in the same situation, as when formerly, the notion of gravitation came into his mind. It was occasion'd by the fall of an apple, as he sat in a contemplative mood. Why should that apple always descend perpendicularly to the ground, thought he to himself. Why should it not go sideways or upwards, but constantly to the earths centre? Assuredly, the reason is, that the earth draws it. There must be a drawing power in the matter: and the sum of the drawing power in the matter of the earth

Figure 1. The author (wearing a light colored jacket distinct from the dark jackets worn by his artillery crew) fires a Civil War cannon to "demonstrate" Newton's point, twisting his body to the left to yank the leather firing cord as instructed by his sergeant. The cameraman is barely visible in the lower right together with the winter hat of the director of the TV film.

> must be in the earths center, not in any side of the earth. Therefore dos this apple fall perpendicularly, or towards the center. If matter thus draws matter, it must be in proportion of its quantity. . . . That there is a power like that we here call gravity, which extends its self thro' the universe.
>
> —W. Stukeley (in his *Memoirs of Sir Isaac Newton's Life*)

Of the several themes that run through this book, Newton's unification of celestial and terrestrial mechanics, which we will come back to in chapter 7, underscores Einstein's amazement, noted in chapter 1, that the universe was comprehensible. What if the planets and the stars were in fact glued to some rotating celestial sphere? What if the computer program[13] that drives the mechanism behind the sphere runs on a logic of its own, random, illogical, and beyond physics? Celestial mechanics would remain the monopoly of deep thinking theologians and philosophers, forever beyond the ken of physicists.

Universal, not parochial

The laws of physics on earth extends to the heavens. They are the same here, there, and everywhere. Not only to the moon and the sun, but far far beyond, throughout the entire universe. We should hope so, that the laws of physics are not like some local ordinances that vary from town to town,[14] from county to county, from state to state, and from country to country. We want the laws of physics to be universal, not parochial.

That the world is comprehensible implies the existence of incorruptible laws of physics, logically interlocked with each other and painstakingly verified by empirical observations. We expect the laws of physics to be universal, rather than the latest whim or fancy of some big chiz at a prestigious university, so that whose theory is correct is a matter of who has a bigger chair and who has more fawning disciples.

Physics inside the stars is the same as physics on earth

Within classical physics, we could see that the motion of the distant planets obeys the same laws of Newtonian mechanics as the motion of the moon, and for that matter, any chunk of rock. But for obvious reasons, physicists felt that an understanding of the inner workings of distant stars would be forever out of reach.

Then, unexpectedly, quantum physics came along.

Why would the study of atoms allow us to look inside stars? Surprise surprise.

Almost incredibly, quantum mechanics with its apparent weirdness often offers us tools to probe where we could never go, such as the interior of stars. The energies and orbits of electrons inside atoms are quantized, that is, they are allowed to assume only certain values specified by quantum physics. Light emerging from the innards of stars on its way out excites the atoms of various elements. A photon, a quantum of light, hits an electron and causes it to jump from one orbit to another orbit with higher energy. By energy conservation, light quanta with the specified energy are absorbed, but not others.

A prism, or the water droplets in the air after a rain, can split incoming white light into a rainbow of colors, in other words, light with different oscillation frequencies.[15] In the same way, ingenious devices (for instance diffraction gratings) were invented[16] to split light in a more controlled and precise manner. By analyzing the spectrum of sunlight and comparing the absorption lines of various gases measured here on earth, physicists could see in detail which frequencies of the light were absorbed as it made its way outward through the sun and so deduced the presence of various atoms in the sun. Indeed, the element helium, named after the Greek sun god Helios, was first discovered in sunlight during the solar eclipse of 1868, and only found on earth much later.

From this point on, the universe, not just the sky, is the limit. The spectra of light emitted by distant stars and galaxies are found in accord with the spectra measured on earth, allowing for the well understood red shift[17] associated with the rotation and movement of the galaxies and the expansion of the universe. Thanks to this faith in the universal rule of physical laws, physicists can not only reach for the stars, but far beyond. Thus was physical cosmology, as in contrast to speculative cosmology, born.

Physicists shocked by the fall of parity

In physics, hope and wishful thinking are not enough, of course, as we have seen repeatedly, and perhaps most emphatically, in recent decades. Physics is based on observational tests.

Throughout the history of physics, long-held beliefs sometimes have to be revised, often to the shock of the physics community. A famous example is the apparently gold plated assertion that the laws of physics do not distinguish between left and right, known as parity. Long held to be "philosophically self evident," it was disproved in 1956 by a landmark experiment performed by the Chinese American physicist Madame C. S. Wu[18] (see figure 2) and collaborators. It turned out, strangely enough, that three of the four fundamental interactions respect parity, which was however violated[19] by the weak interaction. In the aftermath, various other symmetries were also violated. I will come back to this in chapter 7.

The laws of physics were the same eons ago

In everyday life, light certainly seems to get from here to there instantaneously, with infinite speed. But in 1676, the Danish astronomer Ole Rømer[20] showed, by observing the eclipse of the moon Io of Jupiter,[21] that the speed of light (denoted by c much later[22]) was finite, not infinite. (Strangely, Rømer did not push through the calculation, but left his data for the Dutch physicist Christiaan Huygens to use to determine c to be approximately 200,000 kilometers per second, impressively close to the actual value of about 299,792.458 kilometers per second.)

Figure 2. Madame Wu with J. Robert Oppenheimer and her Berkeley advisor Emilio Segrè. Most physicists would agree that Segrè would rank last among the three in terms of contributions to physics, yet he was the only one who received a Nobel prize. In any case, he is certainly far less well known than the other two, particularly after a much acclaimed film in 2023 about Oppenheimer. I am tempted to label this photo, which underlines one of the better known goofs by the Swedish Academy, as "Swedish folly." (I also read recently that a Swedish ophthalmologist expert in geometrical optics was instrumental in blocking the Nobel prize for Einstein.) © 2010 The Regents of the University of California, Lawrence Berkeley National Laboratory.

This crucial fact, that light takes time to get from here to there, allows us to access the physics of eons past. When we look at a distant galaxy, we are seeing it as it was long ago. This fact allows us to gain confidence that the laws of physics have not changed over cosmological time.

For instance, when a gravity wave[23] was first detected in 2016, coming from two black holes embracing eons ago, the fact that physicists could calculate the shape of the wave indicates that Newton's and Einstein's theories[24] of gravity had the exact same form eons ago as they have now, and, we believe, will have eons from now.

Cosmology as we know it would not be possible if the laws of physics change drastically with time. Au contraire, physicists can talk about the early universe with measured confidence. Cosmologists believe that they know exactly what happened within the first few minutes of the Big Bang.[25] In particular, the spectacular agreement of the observed abundances of various elements in the universe with those predicted by Big Bang nucleosynthesis assures us that the laws of physics have not changed substantially for the last ten billion years.

Incidentally, after the discovery of gravity wave, there were expressions of surprise in the popular press that it travels with the same speed c as electromagnetic waves. Actually, as Einstein taught us and as I will elaborate in chapter 6, c is set by the architecture of spacetime, and so is not specific to light. Any massless particle, be it a photon or a graviton, would have to travel at the speed c. So, calling c the speed of light is, strictly speaking, somewhat exclusive.

The laws of physics as distinct from the manifestations of these laws

It is important to distinguish the laws of physics from the manifestations of these laws. The mass of the proton, for example, is not a law of physics, but merely a manifestation of the laws of physics. In the early universe, protons didn't even exist. The temperature was so hot that protons dissociated into their constituent quarks. Yet the laws of physics dictating how quarks bind into protons were the same then as now. As another example, here on earth we could measure (and calculate) the electrical conductivity of copper wire, but inside the sun copper atoms are dissociated, let alone the wires formed of them. Yet the laws of electromagnetism that superintend the collective motion of electrons in the presence of copper ions are assuredly not changed.

Setting limits: upper and lower bounds

As a result of the fall of parity, the social psychology of the physics community changed dramatically after 1956. What was a shock then turned into an easy game. A small subset of physicists started to talk constantly about the possible violations of various "sacred truths," the more sacred the better. The statement "The laws of physics are the same here, there, and everywhere, the same yesterday, today, and tomorrow" is also subject to being disproved.

Physics is however a quantitative science, and it is not enough to just publish a paper saying that "Nyah, the laws of physics have been changing." The standard practice is that an upper limit has to be set, along the following line, "After examining all the evidence, we show that the change in such and such a parameter has to be less than one part in ten million over a period of three billion years."

I certainly feel that those who set these limits are doing useful and necessary work. Notice that I say parameter, not law. People who work in this area typically do not envisage a law of physics suddenly changing, for example, to have the gravitational force changing from an inverse square to an inverse cube law. Rather, they imagine various parameters, such as the mass of the electron m_e and of the proton m_p, changing.[26] For instance, by studying distant quasars, people have concluded that the fractional change in the ratio m_e/m_p equals approximately[27] $(0.26 \pm 0.30) \times 10^{-5}$, which is consistent with zero given the quoted uncertainty.

But what would shock the community is not lots of upper limits (or results consistent with zero), but a single trustworthy lower limit! Perhaps one day, a refereed and published paper will conclude that "After examining all the evidence, we show, with a confidence level of 95%, that the change in such and such parameter must exceed one part in ten million over a period of billion years." The key word here is "exceed."

The practical attitude of most working physicists is something like "Message me when somebody firmly establishes that parameters we believe to be constant have in fact varied. Meanwhile, I am not paying the slightest attention."

Bottom line: The evidence in favor of the universal applicability of physics as we know it has always been, and still is, extremely good.

Notes

[1] S. Chandrasekhar, *Newton's Principia for the Common Reader*, Oxford, 1995.

[2] In this regard, I prefer the Darwinian explanation. Long ago, some apples fell down, others fell up, but the latter did not get to reproduce. Aristotle, living in Greece, talked about rocks, not apples.

[3] The founding father of physics filled notebooks expressing his hatred for his mother for remarrying. As is fairly well known, Newton himself never married. By all accounts, Newton had a chip on his shoulder probably due to his bitter childhood and youth and was not an easygoing fellow.

[4] Albert Einstein "On the Method of Theoretical Physics," lecture delivered at Oxford on June 10, 1933.

[5] In the 1950s and 1960s some leading particle physicists declared that spin was an unnecessary complication, a view that turned out to be more wrong than wrong. See *QFT ASAP*, page 194.

[6] J. Harte, *Consider a Spherical Cow*, University Science Books, 1988.

[7] According to several historical sources, at first Newton did not get the same acceleration experienced by the apple and the moon for what we would now regard as a "trivial" reason: the values for the radius of the earth and for the distance between the earth and the moon were off from their true values. W. Weston (in his *Memoirs* published in 1749) said, "Upon this Disappointment, which made Sir Isaac suspect that this Power was partly that of Gravity, and partly that of Cartesius's Vortices, he threw aside the Paper of his Calculation and went to other Studies." I feel badly for Newton that instead of a moment of triumph, he experienced a major disappointment. But I think I speak for all physicists that we are happy that Descartes's theory of vortices had long been ruled out.

[8] A simple calculation shows that the moon is in fact zipping by at about a thousand meters per second. The fastest bullet with present technology comes in at about eight hundred meters per second.

[9] You had one more shot, muttered under your breath. "If after your father's death, your mother did not marry this rich creep, you would be just like me, pushing a cart in the mud. Hoity toity eh, after all, you are the first one in your family to know how to write your own name!"

[10] At least to some Greeks. The evidence comes from the changing shape of the moon once you realize that light and shadow are at work, and from the hull of a sailing ship disappearing over the horizon before the mast.

[11] For a cartoon illustration of what Newton had in mind, see my chapter in *E = Einstein: His Life, His Thought, and His Influence on Our Culture*, edited by D. Goldsmith and M. Bartusiak, Sterling Publishing, 2007.

[12] The plague years, when Newton did his famous work on gravity at home, were 1665 and 1666, and so Stukeley was talking about a dinner sixty years after the famous event.

[13] Skip the next seventeen prime numbers and move to the one after that. Add all the digits in some arbitrary base that changes in time, multiply all the digits, and use the resulting numbers to determine the direction and speed of rotation in the next instant, but use a smoothing algorithm to conceal the otherwise jerky motion from those nosy humans. I am sure that my friends in computer science could dream up more elaborate schemes.

[14] See a footnote in chapter 9 about the great physicist Ludwig Boltzmann.

[15] Light with different frequencies is seen to have different colors.

[16] The Philadelphia inventor David Rittenhouse made an early device in 1785 by stringing hairs between two finely threaded screws.

[17] Due to Doppler shift, as is familiar in everyday life from the change in frequency of a police car siren as it approaches us and as it recedes from us.

[18] See, for example, "Universe in Reverse: The Queen of Physics Chien-Shiung Wu," https://lough-and-behold.medium.com/universe-in-reverse-the-queen-of-physics-chien-shiung -wu-1ba8f6b3cfff.

[19] The Nobel prize for this shocking revelation went to the Chinese American theorists T. D. Lee and C. N. Yang, who suggested various experiments to test whether parity was violated. It has long been, and continues to be, debated among physicists why the Swedish Academy excluded Madame Wu, considering that by the "rule of three" they could have easily included her. One argument is that another experiment confirming parity violation was performed shortly afterward, but that a second experiment was done has not in other cases excluded the first experiment from the prize. The general suspicion, which I also subscribe to, is simple misogyny prevalent at that time.

[20] The first time I gave a talk at the Niels Bohr Institute in Copenhagen, in the lecture room made famous through photos of the early days of quantum physics, I noticed portraits of two distinguished Danes looking down upon me. I recognized one of them as Niels Bohr, of course, but I didn't know who the other was. During my talk, I mentioned that cosmology was made

possible by the epoch making measurement of c by Rømer. After my talk, I idly asked one of the local physicists who was the person who was not Bohr. He laughed and said, "The very same Rømer you paid tribute to in your talk!" Contrary to what some physicists might think, the family name Rømer, especially when the name is written without the Danish slash, has nothing to do with the Italian city, but was chosen by Rømer's father, one Christen Pedersen, to distinguish himself from the numerous other Pedersens in Denmark and to indicate that the family was from the Danish island of Rømø.

[21] Like a coquettish (Spanish or Chinese) beauty hesitantly showing her face behind a fan, Io would disappear and then reappear by going behind Jupiter. The duration of the eclipse was well measured and found to vary according to the earth's distance to Jupiter. This variation was correctly theorized by Giovanni Cassini, the brilliant Italian astronomer working in Paris and Rømer's mentor, to be due to the finite speed of light. Curiously, Cassini did not pursue his own idea.

[22] Not until the early 20th century, post-Einstein by the way. The letter c stands for celeritas.

[23] I know that this is idiosyncratic usage, but I prefer it to the standard gravitational wave for reasons I stated on page 5 of my book *On Gravity*.

[24] I invoke two great names here. While gravity wave and black hole are specific to Einstein's theory, the orbital dynamics of the two black holes orbiting each other, which determines the frequency of the gravity wave etc., is determined by Newton's theory. It is only at the last stage of the embrace that Einstein's theory is needed. See, for example, *FbN Physics*, chapter VII.4.

[25] S. Weinberg, *The First Three Minutes*, Basic Books 1988.

[26] Indeed, since the origin of masses is understood nowadays, at least in outline, something like masses changing extremely slowly with time could be accommodated within our theoretical framework without too much stress.

[27] For an overview, see, for example, V. Flambaum and J. Berengut, page 383, in *Proceedings of the Conference in Honour of Murray Gell-Mann's 80th Birthday*, edited by H. Fritzsch and K. K. Phua, World Scientific, 2011. This particular result was reported on page 388.

3
THE WORLD IS QUANTUM

Mysterious and mystifying

> Those who are not shocked when they first come across quantum theory cannot possibly have understood it.[1]

Unless you have been headhunting in the jungle of Papua New Guinea, you have surely heard that the classical world we were born into and love so passionately is merely a surface manifestation of the underlying quantum world. What we commonly think of "reality" in everyday life is but a mathematical approximation[2] to a bizarre quantum "reality." Classical "reality" turns out to be an approximation that holds to a fantastic degree of accuracy when the objects involved are lumbering giants compared to the true actors in the drama of the universe, namely the quarks and their friends, the electron and the neutrinos. The common sense we gleaned and compiled from birth fails utterly in this mysterious and mystifying quantum world.

Two astonishing insights rocked 20th century physics: that the world is relativistic and that the world is quantum. While learning that space and time can be unified and bent and that gravity is due to the curvature of spacetime blew my mind, still that cannot compare with the head spinning knowledge that there is an unseen world underlying our world and from which our world springs. But who is comparing?

The scope of our discussion

I would like to say immediately that, considering how physicists continue to find the quantum world so puzzling compared to curved spacetime, there is simply no way for me to explain much about quantum physics in a single chapter in a popular book. I am fortunate to be among those who have written successful popular books as well as textbooks and so

I can say categorically that this is for sure not a textbook. My strategy here is to bring out some paradoxical aspects of quantum physics by talking about the Stern-Gerlach experiment, which did much to crack open the atomic domain. This would then allow me to describe the issues the quantum experimenters who won the physics Nobel prize in 2022 sought to address. I have no choice but to sweep various subtle issues under the rug.

I suspect that the typical reader of this book already knows something (or even a great deal) about quantum physics, given that it is bandied about on the internet incessantly.[3] Yet the potential readers of this book presumably span an entire range. For those who are hazy about the whole subject, I feel compelled, if only to fix our minds into the quantum mindset, to give a lightning overview of how the quantum world was discovered. The story is more exciting than any adventure movies I have ever seen about discovering some lost world in the jungle.

A lightning review of history

Many roads led to the quantum world. Historically, a multitude of phenomena pointed to the failure of classical physics. For our purpose here, let me mention just a couple of them. By the early days of the last century, the atom was known to consist of a bunch of negatively charged particles, soon named electrons, orbiting around a positively charged nucleus. According to the classical theory of electromagnetism, charged particles, unless they move in a straight line at a constant speed, radiate electromagnetic waves, which carry away energy. Indeed, any number of household devices are based on this principle, from the lightbulb to the microwave oven. Thus, physicists were immediately faced with the difficulty of explaining why in an atom the electrons do not radiate away their energy of motion, spiral inward, and crash into the nucleus.

Meanwhile, excited atoms (for instance in a hot gas) do radiate electromagnetic waves, but curiously, only at certain frequencies. In other experiments, white light passing through a gas of atoms has some of its frequencies absorbed. Which frequencies are radiated or absorbed is characteristic of the type of atoms. (Indeed, that was how helium was first discovered in the sun, showing that the laws of physics are the same on earth and far away from earth, as was mentioned in chapter 2.)

These puzzling observations sketched here could be understood if an electron in an atom were allowed to have only certain energies. An electron could exist only in certain "energy levels." Its energy is said to be quantized.

When atoms collide in a hot gas, some of the electrons are theorized to be kicked up to some higher energy level. An electron in a higher energy level could jump to a lower energy level, in a process known as a quantum leap (which happily has become part of the English language), and thus loses energy equal to the difference in energy between the two levels. This energy is carried away by a photon. (Other lines of investigation showed that electromagnetic waves consist of particles of light called photons, just as matter consists of atoms.) Similarly, the absorption of a photon carrying the right amount of energy by an electron could kick it from a lower level to the higher level. If we could somehow relate the energy carried by a photon to the frequency of the electromagnetic wave it is associated with, then we would have explained why atoms radiate and absorb only certain characteristic frequencies. This entire endeavor, known as atomic spectroscopy, proved invaluable for the study of atomic structures.

If electrons could only exist in certain energy levels, and if there is a level with the least amount of energy (called the ground state), then as an electron leaps from one level to another with lower energy, it will soon find itself in the lowest energy level. It can't go any lower. Ergo, it cannot crash into the nucleus. That explains the stability of atoms.[4]

A hundred and some years later, I am still astounded by the bold originality of the founding fathers of quantum physics in how they resolved these problems. How did these guys do it? How did they come up with such weird counterintuitive stuff?

A partial answer is that there was an embarrassing richness of experiments. The discovery of quantum mechanics is a beautiful example of the enormous intellectual resonance that could result when experiment and theory march forward hand in hand.

What I described here[5] is known as the Bohr model, named for the heavyweight Danish physicist. With some refinements added, it accounts for most of what was known in the 1920s.[6]

Planck's constant

One sentence in the previous section is less innocuous than it looks. Can you spot it? It starts with "If we could somehow relate the energy carried by a photon to the frequency of the electromagnetic wave." Frequency is defined as the number of repetitions or cycles per second, and so frequency, as a physical quantity measured by some number per unit of time, has dimension of 1 divided by time, which we will write as $1/T$. On the other hand, energy is measured in units of energy. So how could you possibly relate two quantities measured in different units?

Max Planck, the great German physicist who launched quantum theory,[7] was thus forced to introduce a hitherto unknown constant of Nature, now known as Planck's constant[8] (naturally) and written as $\hbar$,* to relate energy E to frequency ω via $E = \hbar\omega$. In other words, $\hbar$ is fixed, in modern language, as the ratio of the energy and the oscillation frequency of a photon, the quantum particle of light. Clearly, $\hbar$, being equal to E/ω, has dimension of energy multiplied by time, that is, the dimension of ET, since frequency is the inverse of time, as noted above.

Recall from high school physics that a particle with mass m moving at a speed v has a kinetic energy of $\frac{1}{2}mv^2$. Thus, in some system of units devised by French revolutionaries, a mass of 1 gram moving at 1 centimeter per second has an energy of $\frac{1}{2}$ gram centimeter squared per second squared. Multiplying this by one second, we see that Planck's constant has dimension of gram centimeter squared per second. Doesn't sound like the dimension of anything you would encounter in your everyday life. Or does it?

Recall also from high school physics that an object of mass m moving with velocity v around a circle of radius r has angular momentum mvr, namely mass times velocity times radius. This expression simply conveys the intuitive fact that an object's angular momentum is correspondingly larger if the object is more massive, moves faster, or moves in a larger circle. Thus, the unit of angular momentum is gram times centimeter per second times centimeter, that is, gram centimeter squared per second. Looks familiar?

Yes, the mysterious constant $\hbar$ introduced by Planck when he launched quantum physics has the dimension of angular momentum and thus provides a natural unit for measuring angular momentum. As will be noted shortly, physicists soon realized that angular momentum is quantized in units of $\hbar$, that is, angular momentum could only take on values equal to an integer times $\hbar$.

The measured value of $\hbar$, about 10^{-27} gram centimeter squared per second, so preposterously teeny compared to anything in everyday

---

*This peculiar symbol $\hbar$ consists of the lowercase letter h with a slash through it, and is pronounced by physicists as "h bar."

experience, indicates why physicists discovered $\hbar$ and quantum physics only after they started exploring the microscopic world of atoms. Consider that a mass of 1 gram moving around in a circle of radius 1 centimeter with the speed of 1 centimeter per second would have angular momentum of $10^{27}\hbar$! Since a million is 10^6, a billion 10^9, and a trillion 10^{12}, this tiny mass orbiting around has an angular momentum a thousand trillion trillion times larger than the fundamental unit the universe reckons angular momentum with. This is why everyday "reality" as we experience it appears to be totally continuous: as the angular momentum of an everyday object changed from $(10^{27} - 2)\hbar$ to $(10^{27} - 1)\hbar$ and then to $10^{27}\hbar$, to $(10^{27} + 1)\hbar$, and so on, we would barely notice that it is even changing, let alone by a discrete jump in each of the steps along the way. It is only when the angular momentum becomes say $(10^{27} + 10^{25})\hbar$ (after 10^{25} steps!) that an experimentalist with some fancy equipment might have noticed that it has changed by a measly 1%.

Bohr's wild and erroneous guess

In the struggle to understand the atom, Niels Bohr observed crucially[9] that Planck's constant $\hbar$ has the dimension of angular momentum, as we just noted. In classical physics, an object could spin with any amount of angular momentum, just because 10^{27} is such a huge number, as mentioned earlier. In contrast, in quantum physics, angular momentum, measured in unit of $\hbar$, is quantized to equal an integer, either positive or negative: 0, $\pm 1, \pm 2, \pm 3, \cdots$. In the quantum world, angular momentum could only be equal to an integer times $\hbar$. Hey, that's why we call it quantum physics.

By the mid 1920s, after almost three decades struggling with the mysteries of the atomic domain, physicists finally understood that angular momentum is quantized in the manner I just described, and using that fact, they began to master atomic spectroscopy in broad outline. In fact, Bohr's breakthrough, which cracked atomic physics wide open, amounts to guessing[10] that the electron orbiting the proton in a hydrogen atom in its lowest energy state has angular momentum equal to one unit of $\hbar$.

In the glare of hindsight, every student of quantum physics nowadays knows that Bohr's guess is wrong, that the orbital angular momentum in that state is actually zero. But through a remarkable stroke of luck, Bohr's erroneous guess led to the correct answer for the measured energy of the hydrogen atom in its lowest energy state.

The electron spins with only half the amount of angular momentum quantum mechanics allegedly allows

Yet there remained many puzzling aspects to atomic spectroscopy.[11]

In late 1925, the intriguing idea that the electron spins occurred to two students in Leiden, the Netherlands. The story[12] of how these two young people, George Uhlenbeck and Samuel Goudsmit,[13] discovered the spinning electron, against a tide of skepticism from their elders, is still told with relish by physicists. The introduction of electron spin swept away many of the remaining puzzles in atomic spectroscopy, and provided one of the keys to finally unlocking the laws of quantum physics as we know it.

But surprise surprise. The two students, who had yet to receive their doctorates,[14] found that in order for their proposal to fit atomic spectroscopy, the angular momentum associated with the spinning electron has to equal $\pm \frac{1}{2}$. Not an integer, but a half integer!

The shock was not merely that the electron spins, but that it spins with half of the amount of angular momentum previously thought to be allowed by quantum mechanics. The weird thing is that the angular momentum of an electron moving around is quantized in units of $\hbar$, but the angular momentum of an electron spinning can have the value of $\pm \frac{1}{2}\hbar$. Indeed, this fact expresses a profound essence of the 3-dimensional space we were born into.

The half manifests itself in bizarre ways, as exemplified by the following fact about rotations in the quantum world. In everyday life, if we rotate an object by 360°, we would bring it back to its original configuration. But we have to rotate an electron by $2 \times 360° = 720°$ in order to bring it back to its original configuration! $2 \times \frac{1}{2} = 1$! Get it? Physicists say that the electron spin double covers[†] rotation.

Incidentally, this bizarre half integral quantization continues to bedevil physics students, and many still fail to wrap their heads around it.

A prominent physicist at the time likened the discovery of particles with half the quantized unit of angular momentum to the chance encounter of a hiherto unknown tribe of headhunters in the deepest recess of the jungle. But it was more than that. It was as if these headhunters had only half a head: one ear, one eye, one nostril, and one cheek.

[†]Curiously, mathematicians, notably Eliè Cartan, had already discovered this possibility of double covering the rotation group. See *Group Nut*, chapter IV.5.

Three young guys, one of whom told his idea to the wrong "old" guy and ended up the loser

I was so dumb when I was young.

—Wolfgang Pauli

Unbeknownst to Goudsmit and Uhlenbeck, a young German physicist named Ralph Kronig, who was only twenty-one in 1925, had the same idea a few months earlier. Unusually, in an era when many American physicists later to become eminent went to Europe to study (as everyone who has seen the movie *Oppenheimer* now knows), Kronig went to Columbia University to study[15] physics. He was subsequently advised by a visiting Paul Ehrenfest (who was, small world, the professor of Goudsmit and Uhlenbeck in the Netherlands; more about him later) to go back to Europe. "Young man, that's where the action is in physics!"

Before publishing his idea that the electron spins, Kronig made the mistake[16] of consulting Wolfgang Pauli, who was legendary for his sharp tongue and critical insight. In classical physics, the electron was estimated to have a tiny radius[17] for use in various contexts. Pauli pointed out that, given the tiny radius of the electron, if the electron were to spin, the surface of the electron would have to move much faster than the speed of light.[18]

Having his idea dismissed as contradictory to what Einstein had decreed, the devastated Kronig withdrew his paper[19] from publication. Later, Pauli would blame himself, saying, "Ich war so dumm wenn Ich jung war." (I intentionally wrote this in the original German just to show readers who don't know the language how close it could be to English. Hint: j corresponds to y in English.)

In contrast, the affable[20] and encouraging Paul Ehrenfest liked Uhlenbeck's and Goudsmit's idea of the spinning electron, and offered to arrange for publication of their paper, which by the way was a one page letter to the editor. Meanwhile, Lorentz, the grand old man of Dutch physics (more about him in chapter 6) already in retirement living in Haarlem, visited Leiden. When the young Uhlenbeck told the old Lorentz about electron spin, the latter also pointed out that the surface of the electron would be moving faster than the speed of light.[21] Shattered, Uhlenbeck rushed to Ehrenfest to withdraw the paper, but Ehrenfest told him that it was too late, adding, "You and Goudsmit are both young enough to be able to afford a stupidity."[22]

Interestingly, Pauli and Ehrenfest, who were both Austrian by the way, had such diametrically opposite personalities and attitudes toward younger people. One possible "explanation": Ehrenfest was twenty years older than Pauli and belonged to Einstein's generation, while Pauli, a child prodigy, was actually not that old.

Indeed, Pauli, already a theoretical physicist of the first rank at the time, was the same age as Uhlenbeck, who was, in turn, two and four years older than Goudsmit and Kronig respectively. The reason for Uhlenbeck's age was that students in the Netherlands in that era were not allowed to study physics unless they were versed in Latin and Greek. I understand that this was effectively discrimination based on social class. Young Uhlenbeck, unschooled in the classical languages, was thus forced to study chemical engineering, which he hated, at an institute of technology. Fortunately for him, the law was soon changed, and he transferred to the university in Leiden in 1919, the center of Dutch physics at the time. To earn money, he had to teach math in a high school, and then in 1922, he took a job in Rome tutoring the son of the Dutch ambassador and almost abandoned physics. Due to his involuntary detour into chemical engineering, Uhlenbeck had to be tutored by the younger Goudsmit, who had already accomplished much in atomic spectroscopy.

Pauli was one of the most colorful major figures in theoretical physics, and to this very day physicists delight in stories about him. He once said that it would be impossible to draw a caricature of him because he, Pauli, was already a caricature.

The spinning electron and Bohr's train journey

If you take an electrical device apart, you would often find a coil of wire wrapped around an iron bar. When a current is turned on, electric charges cruising through the coil transform the bar into effectively a magnet. Electric charges moving around in circles act like a magnet. See figure 1(a).

Since the electron carries an electric charge, a spinning electron also acts like a tiny magnet: it is said to have a magnetic moment, denoted by $\vec{\mu}$ and traditionally represented by an arrow pointing in the direction of the magnet, as in figure 1(b). In a magnetic field, denoted by $\vec{B}$, a magnetism would precess, as shown in figure 1(c), in complete analogy with a spinning top precessing in the earth's gravitational field (which corresponds to the magnetic field $\vec{B}$).

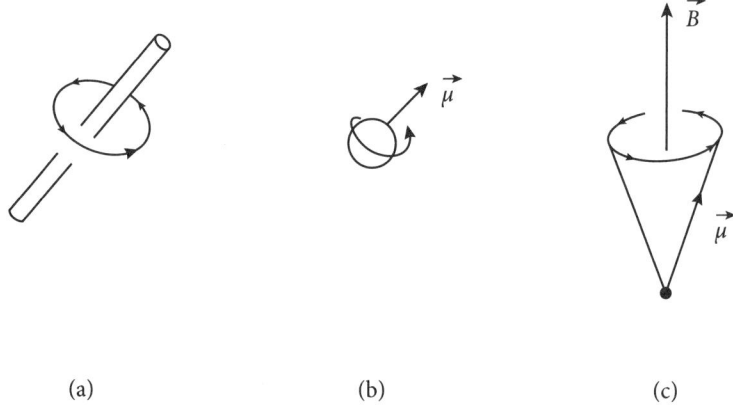

Figure 1. (a) An electric current going around an iron rod turns it into a magnet. (b) A spinning electron acts like a tiny magnet. Its magnetic moment $\vec{\mu}$ is represented by an arrow pointing in the direction of the magnet. (c) The magnetic moment of an electron precessing around a magnetic field $\vec{B}$, just like a child's spinning top precessing around the earth's gravitational field.

In December 1925, Niels Bohr went by train from Copenhagen to Leiden to celebrate the 50th anniversary of Lorentz's doctorate. Bohr's train ride has since become part of the legend of quantum physics.[23] When the train stopped in Hamburg, Wolfgang Pauli and Otto Stern (more about him later) showed up at the station to ask Bohr what he thought of the spinning electron. Bohr said that the idea was very interesting, which was allegedly his polite way of referring to ideas that he did not believe in. When he arrived in Leiden, he was met at the station by Ehrenfest and Einstein, who wanted to know what he thought of the spinning electron. Bohr said that the idea was very, very interesting.

After discussions with Ehrenfest and Einstein, however, Bohr was convinced that the idea was correct and urged Goudsmit and Uhlenbeck to publish a more detailed paper, to which he appended an approving comment. After the celebration in Leiden, Bohr traveled to Göttingen where he was met by Werner Heisenberg and Pascal Jordan, who asked him what he thought of the spinning electron. Bohr replied that it was a major advance. Then the train stopped in Berlin. Well, Pauli had traveled all the way there from Hamburg just to ask Bohr, once again, what he thought of the spinning electron now, after his visit to Leiden. Bohr replied that it was a tremendous idea. Pauli sighed, "Yet another Copenhagen heresy!"[24]

After Bohr got home to Copenhagen, he wrote to Ehrenfest that he had become "a prophet of the electron magnet gospel."

That the electron acts like a tiny magnet underlies much of modern technology and may even provide a basis for the yet-to-be-realized quantum
computer.‡ But for physicists, it provided one more key for unlocking the
quantum world. And so we turn to the story of the famous Stern-Gerlach
experiment.

A bizarre scarf factory employing elves to sort

To explain the Stern-Gerlach experiment, I resort to the following
approach. I will first describe an analogous situation in everyday life
and show that the observations corresponding to what Stern and Gerlach saw make no sense. Hardly news to the readers of this book! Quantum
physics famously goes against common sense. My task here is to show you
how much it goes against common sense.

Imagine a scarf factory that makes only red scarves and blue scarves.
The output of the factory is conveyed on a belt into a sorting room. We
cannot look inside the enclosed room but we are told that an elf inside
sorts the scarves according to color and places them on two belts coming
out of that room, one belt carrying only red scarves, the other only blue
scarves. See figure 2(a). For some reason, perhaps because the factory is
located in a so-called blue state in the United States, the red scarves are
dumped into a truck and taken away, presumably to a red state.[25] (To make
sure that our elf did not make a mistake, we can double check by feeding
the stream of blue scarves into another room where another elf is also told
to sort by color. Sure enough, only blue scarves come out. By the way, this
extra "quality check" process is not shown in the figure 2(a); you could
draw it on the right of figure 2(a) if you wish.)

Thus far, there is nothing surprising in the slightest about the story. But
I do ask you to please remember that all the scarves we have are blue. The
red scarves have all been dumped into a truck and carried away.

Now we feed the stream of blue scarves into a room with an elf who is
totally color blind (or perhaps the room is simply kept in total darkness)
and who is instructed to use his sense of touch to separate the scarves into
silk and woolen scarves. Again, two conveyor belts emerge, one carrying
only silk scarves, the other only woolen scarves. The woolen scarves are
taken away in trucks. This is shown in figure 2(b).

---

‡An electron spinning up and down, corresponding to the bits 0 and 1 in classical computing,
is now called a qubit in quantum computing.

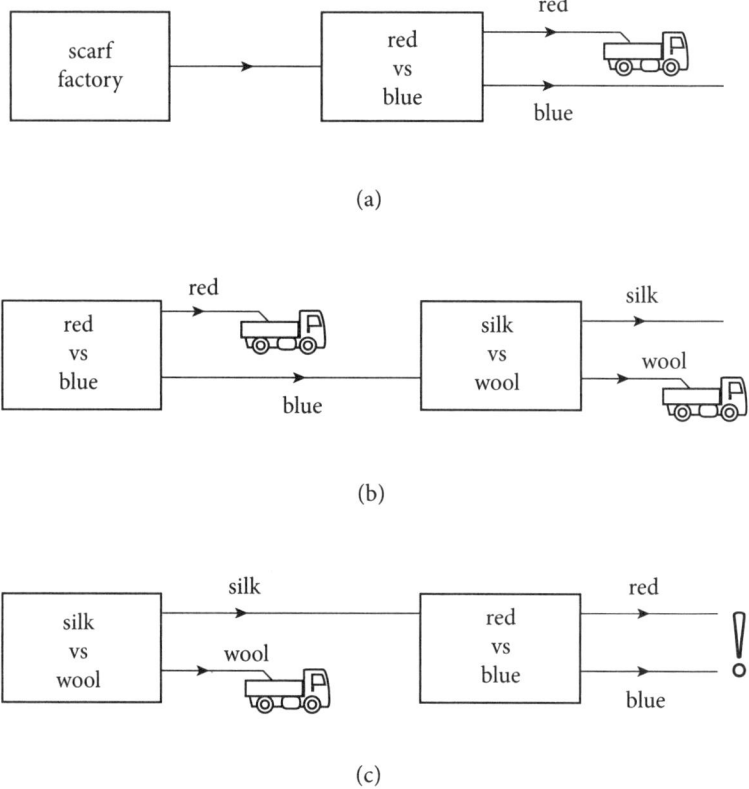

Figure 2. A strange scarf factory described in the text. (a) The scarves are sent into a room labeled "red versus blue" and sorted by an elf into two streams, with the red scarves trucked away. (b) The stream of blue scarves coming out of the sorting room in (a) is sent into a room labeled "silk versus wool" and sorted into two streams, silk and wool, with the woolen scarves trucked away. Note that the left side of (b) is just the right side of (a) repeated for your viewing convenience. (c) The stream of blue silk scarves coming out of the sorting room in (b) are sent into another room labeled "red versus blue" and sorted. Surprise surprise, two streams come out, with equal number of red and blue scarves. Where did the red scarves come from? Weren't they all trucked away?

What is amazing about that? Absolutely nothing.

To underline that absolutely nothing strange or amazing has occurred up to this point in the story, let us imagine you the reader as an eccentric American billionaire. You could set up a scarf factory, hire some elves, buy some trucks, and arrange everything to come out exactly as described. You could even make a (rather dull) movie out of this without resorting to any computer generated special effects.

Next, we feed the stream of silk scarves into a sorting room with an elf capable of separating red scarves from blue scarves. This is depicted in

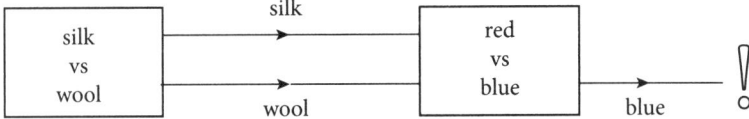

Figure 3. Suppose that in figure 2(b) the truck hired to take the woolen scarves away did not show up. So instead of what was depicted in figure 2, both streams, one consisting of silk scarves, the other of woolen scarves, are sent into a room labeled "red versus blue". Surprise surprise, only blue scarves come out! You might have noticed that this is just figure 2(b) run in reverse.

figure 2(c). Guess what? Surprise surprise, two streams of scarves emerge, with one belt carrying red scarves, the other carrying blue scarves. Equal number of red and blue scarves coming out!

This totally contradicts common sense! In "real life," the red scarves had all been trucked away. How in the blessed world can they reappear?

Here is a variant that is perhaps just as astonishing. Suppose in figure 2(b) the trucks did not show up to carry the woolen scarves away. This is shown on the left of figure 3. So, in this alternate version, the blue scarves coming on the right of figure 2(a) are sent into a room where an elf sorts them by touch into silk and woolen scarves. Thus, two streams, one of silk scarves and another of woolen scarves, come out of the sorting room. We then allow both streams to go into a sorting room operated by the elf with color vision but without a sense of touch (or perhaps because the scarves are wrapped in thick but translucent plastic bags). Do you want to pause for a moment to think about the scarves that would emerge?

Yes, only blue scarves!

What is "reality"?

This is the sort of quantum scenarios that disturbed Einstein deeply, down to his understanding of reality itself. In everyday life, the color and texture of a scarf is part of what we call reality. But in quantum physics the color of a scarf is not determined until you "measure" it (that is, until you look at it). Similarly, whether a scarf is silk or woolen is not determined until you measure it, that is, perform an experiment on it, by touching for example. The Israeli physicist Asher Peres[26] proposes the catchy slogan "Unperformed experiments have no results." But quantum physics is more than this slogan could capture. After measuring the color of the scarf to be blue, we could decide to measure whether it is a silk or a wool scarf. But as soon as we do that, we lose the information about the color. A subsequent measurement could find that it is actually red. Not only do unperformed experiments have no results, but subsequent experiments could shuffle and undo the results of an earlier measurement.

This by the way is the basis of all that apparently crazy talk about Schrödinger's cat being both dead and alive.[27] Perhaps a good place for a rant? I have never liked Schrödinger's cat, which has by now even become a meme for cartoonists. I think that it confuses more than illuminates. Besides, it raises the ire of cat lovers everywhere and furthers the negative image of theoretical physicists as Cruella incarnate.

Stern advice: on a cold winter morning, better to stay in bed and think about physics

Needless to say, the actual experiment was not done with scarves, but with electrons. Over the decades, I have come to classify introductory textbooks on quantum mechanics by whether or not they start[28] with the Stern-Gerlach experiment, and over the decades, the movement has apparently been toward starting with the Stern-Gerlach experiment.

Otto Stern, after serving in World War I, became Max Born's assistant at the Institute for Theoretical Physics in Frankfurt, Germany. As he later recalled, he conceived of the famous experiment one winter morning in late 1921 when he decided that it was too cold to get out of bed. The experiment, performed[29] in 1922 with the help of Walther Gerlach, is often cited in textbooks as a milestone for discovering the spinning electron. But the attentive reader keeping track of dates knows that that can't be correct since the Goudsmit and Uhlenbeck paper proposing the spinning electron wasn't published till early 1926. Stern actually had quite a different goal in mind: he wanted to prove that Bohr's atomic model was wrong. In fact, he and Max von Laue (later a Nobel laureate, just like Stern) had sworn an oath that they would both quit physics if Bohr were right.[30] Serendipity! Trying to do one thing, and discovering something else, as often happens in physics.

The magnetic moment of the electron interacting with a magnetic field

Now I have to give you some physics background. Physics is blessed or burdened, depending on your taste, with a load of peculiar terms. A magnetic moment pointing in the direction of a magnetic field is said to be pointing up, while a magnetism pointing in the direction opposite to that

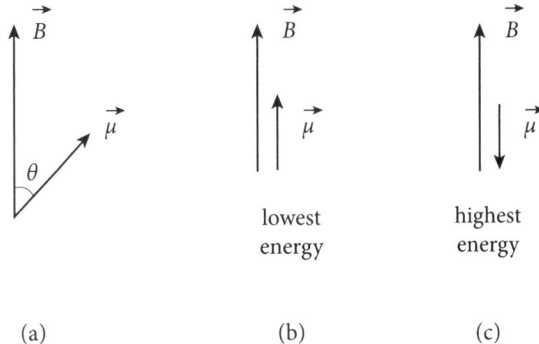

Figure 4. (a) The energy between a magnetic moment $\vec{\mu}$ and a magnetic field $\vec{B}$ depends on the angle θ between them, and is lowest (b) when $\theta = 0°$ and highest (c) when $\theta = 180°$. In other words, the magnetic moment wants to point in the direction of the magnetic field $\vec{B}$.

of a magnetic field is said to be pointing down. (The up and down here have nothing to do with the direction hardwired into our brain by eons of evolving in the earth's gravitational field. It's just a convenient way[31] of labeling two opposite directions. The masses of the elementary particles, such as the electron, are so tiny that the gravitational force acting on an individual electron is utterly negligible.)

In classical physics, the angle θ between a magnetic field and a magnetic moment is free to vary continuously from $0°$ to $180°$. See figure 4. The energy[32] between a magnetic moment $\vec{\mu}$ and a magnetic field $\vec{B}$ (note that they are both vectors, namely quantities with a direction) is lowest when $\theta = 0°$ and highest when $\theta = 180°$, with values in between these two extremes for θ between $0°$ to $180°$. In plain English, the energy is lowest when the magnetic moment and the magnetic field point in the same direction, highest when they point in opposite directions, with values in between. So naturally the magnetic moment, if it is allowed to pivot, wants to point along the magnetic field.[§]

We need to go on yet another technical excursion. Stern couldn't simply put an electron in a magnetic field to see if it has a magnetic moment. In 1921, it was certainly not technically feasible to isolate a single electron. In any case it is not practical to put a single electron in a magnetic field. Since the electron is charged, it would be swept away by any stray electric field.

[§]This imperative to lower one's energy is of course universal in physics, manifest for instance in water flowing downhill and in couch potatoes lying down.

The experiment envisaged by Stern could only be done with electrically neutral atoms. Stern and Gerlach chose silver atoms.

Why silver? Well, electrons like to pair up and cancel each other's magnetic effect (as we now know, by spinning in opposite directions). The silver atom has 47 electrons and after 46 electrons are paired up, the lone leftover electron[33] endows the silver atom with a magnetic moment.

In 1921, nobody dreamed that the electron could be spinning, and so physicists thought that any magnetic moment it may possess must be due to its orbital motion, that is, due to its motion through space around a circular orbit. In Bohr's theory, the orbital angular momentum of this lone electron in a silver atom is quantized to be $+1$, 0, and -1, as was mentioned earlier. In other words, the magnetism associated with the lone electron's orbital motion in a silver atom can point in only three possible directions: along the direction of the magnetic field imposed by the experimentalist, perpendicular to that direction, or against that direction. (At the time, this was known as space quantization, a term that mercifully has since been swept into the dustbin of history. A better term might be directional quantization.)

Sorting electrons by using an inhomogeneous magnetic field

Stern wanted to disprove Bohr's "space quantization," and so he had to think of a way to sort the atoms according to the directions their magnetic moments point in. Thus far, we have been talking about constant or homogeneous magnetic fields. Let us figure out what would happen in an inhomogeneous magnetic field, in particular, a magnetic field that gets stronger as we move along the direction it is pointing in, as indicated in figure 5. For the sake of narrative definiteness, call that direction "up." (You now see why physicists favor this terminology of up and down even when gravity is totally irrelevant.)

Well, an atom with its magnetic moment pointing up would want to lower its already low energy even more by moving to where the magnetic field is strongest, that is, by moving in the direction of the magnetic field. In contrast, an atom with its magnetism pointing down would want to lower its high energy by moving to where the magnetic field is the weakest. (Recall what we said earlier about water flowing downhill?) Thus, an inhomogeneous magnetic field would effectively push the atoms with

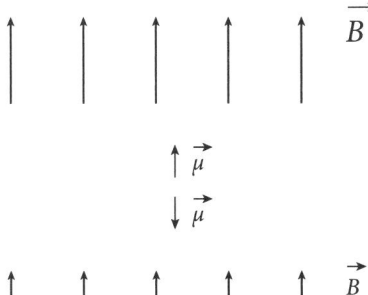

Figure 5. An inhomogeneous magnetic field, in particular, a magnetic field that gets stronger as we move along the direction it is pointing in, as indicated symbolically by the larger arrow and a bigger letter B. An atom with its magnetic moment pointing up would want to move to where the magnetic field is strongest. To the contrary, an atom with its magnetic moment pointing down would want to move in the opposite direction. That atoms in an inhomogeneous magnetic field would migrate and separate according to which way their magnetic moment is pointing in, due to their wanting to lower their energies, is similar to water flowing downhill.

their magnetic moments pointing up and down in opposite directions, thus sorting them.

Stern proposed to heat up a vapor of silver atoms in an oven. When a tiny hole is opened on the side of the oven, a stream of silver atoms will escape, forming an atomic beam. (This basic method of producing an atomic or a molecular beam is still routinely used nowadays in all kinds of research and has given rise to a wealth of technological applications. The spiritual descendants of Stern-Gerlach experiment, all exploiting quantum jumps between "space quantized" states, include nuclear magnetic resonance, optical pumping, the laser, atomic clocks, Lamb shift, the electron's anomalous magnetic moment, probes for nuclei, for proteins, and for galaxies, devices to image bodies and brains, to perform eye surgery, to read data from compact discs, and to scan the human genome.[34] Are you impressed, or are you impressed?) The beam of silver atoms is then passed through an inhomogeneous magnetic field and deposited on a screen. See figure 6.

A difficult experiment, a comedy of errors, and a lack of money

Stern looked for someone experienced with inhomogeneous magnetic fields and recruited Gerlach to perform the experiment with him. I have great admiration for these early quantum pioneers. I think that even a

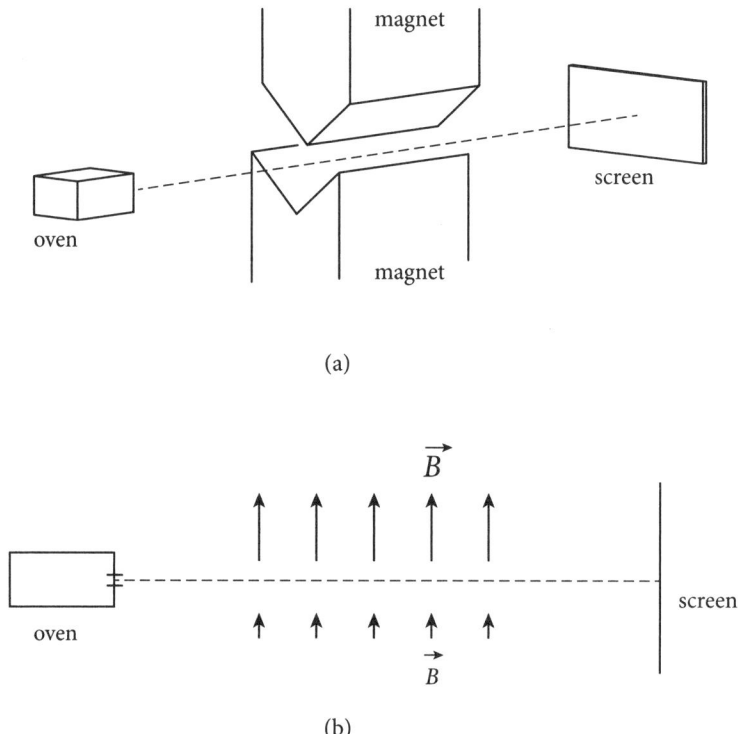

Figure 6. (a) Theorist conception of the Stern-Gerlach experiment. A beam of silver atoms (indicated by the dotted line) emerges from an oven, passes between the two poles of a magnet, and gets deposited on a screen. (b) Schematic side view: The inhomogeneous magnetic field is shown to be stronger above the beam and weaker below the beam. An atom with its magnetic moment pointing up will follow a trajectory bending upward, while in contrast an atom with its magnetic moment pointing down will follow a trajectory bending downward. Atoms with their magnetic moments pointing at angles between up and down will follow trajectories between these two extremes. Thus, Stern expected the deposit on the screen to form a continuous smudge, as explained in the text.

casual reader could see that this experiment, while conceptually simple, is going to be plagued by technical difficulties. The beam of silver atoms has to pass through the best vacuum that pumps at the time could produce: you certainly don't want those silver atoms to collide with air molecules on their fateful journey from the oven to the screen. The magnets generating the inhomogeneous magnetic field have to be strong enough to separate the atoms according to the direction their magnetisms are pointing in. Stern had to calculate the weakest magnets capable of doing this and then ask Gerlach if they would be available. By the way, the Stern-Gerlach experiment also characterizes how experimental collaborations are formed; the

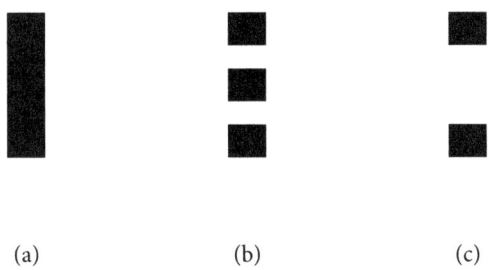

Figure 7. (a) The continuous smudge expected in classical physics. (b) Three smudges expected in quantum physics if the magnetism is due to an orbiting electron with angular momentum $+1$, 0, -1. (c) Two smudges expected in quantum physics if either the magnetism is due to a spinning electron or Bohr's totally erroneous argument actually holds.

person who conceived the experiment has to work with the person with the necessary technical expertise, obviously.

Classically, the escaping silver atoms would emerge from the oven with their magnetisms pointing every which way. So Stern, not believing Bohr's space quantization nonsense, expected those atoms pointing up to land at the top of the screen, those pointing down to land at the bottom, and those with angles between $0°$ to $180°$ land in between. He expected to see a continuous smudge on the screen, as would be predicted by classical physics.

Nowadays, any student studying quantum physics knows that classical physics does not hold in the atomic domain, and would say that three dot-like smudges,[35] corresponding to $+1$, 0, and -1, should appear on the screen instead of the continuous smudge of classical physics. But no, to add to the comedy of errors that is said about some of these famous early quantum experiments, the theoretical expectation of the time was also wrong! Bohr had published a paper, based on a completely wrong argument, mercifully now long forgotten, stating that the orbital angular momentum 0 was not allowed, and so he predicted two smudges corresponding to $+1$ and -1 (see figure 7).

Soon after the experiment got under way, money ran out—naturally: Germany was impoverished after World War I. Max Born, the senior theorist of the group, resorted to giving popular lectures on Einsteinian relativity, which the public was tremendously interested in, and charged admission. The experiment continued for a while but again money ran out. Fortunately, Born had a friend who knew a financier named Goldman in New York, and at the friend's suggestion, wrote in desperation to the rich man for help. To everyone's surprise, some much-needed dollars were wired to Frankfurt for the experiment to continue. So Goldman Sachs[36] had a hand in launching quantum physics!

The cheap cigar that saved the Stern-Gerlach experiment

Once the apparatus started working, Stern and Gerlach eagerly examined the screen, and they saw ... nothing! Apparently, the density of silver atoms deposited was too low to be visible.

Lady Luck then favored physics some more. Stern and Gerlach were both fond of smoking cigars. According to this particular physics legend, while Gerlach was fiddling with the experimental apparatus, Stern stared hard at the screen puffing away the whole time. Being young, he could only afford cheap cigars with a high sulfur content. Remarkably, some smudges began to form, but Stern and Gerlach couldn't resolve the smudges. We now know that the sulfur atoms in his cigar smoke reacted with the silver atoms on the screen to form the chemical compound silver sulfide, which happens to be black.[37] Nature is so incredibly kind to physicists!

I kept thinking that in our utopian days some highly paid administrator would pop out of the closet, eject Stern and Gerlach from their lab, and shut down the experiment. Cigars are no longer allowed on campus, let alone in an atomic lab.[38] Perhaps more likely these days at American universities, so many bureaucratic forms have to be filled out that would-be pioneer experimenters like Stern and Gerlach simply give up in disgust.[39]

One more fortunate turn of events! By early 1922, the real life Stern and Gerlach did consider giving up in frustration. By then, Stern had left Frankfurt for an academic position in another city. He and Gerlach met in Göttingen, and after much agonizing discussion, decided to abandon the experiment!

This time, Lady Luck pulled a railroad strike out of her sleeve. Gerlach, stuck and idled in Göttingen, thought about how to improve the experiment and to possibly give it one more try. After returning to Frankfurt, he aligned the magnets better, made some other improvements, and saw two smudges. Not a continuous smudge, not three smudges, but two! He sent Stern a postcard saying that "Old Bohr is right after all!" (By the way, "Old Bohr" was thirty-seven at the time.)

And so Stern and Gerlach, thinking that they had confirmed what we now know is Bohr's nonsensical argument, sent Bohr a famous congratulatory postcard with a photo of the two smudges. Recall that Goudsmit and Uhlenbeck had yet to propose the spinning electron; indeed, in 1922 Uhlenbeck had barely escaped from his career as a chemical engineer and, living in Rome tutoring some rich kid (as I mentioned earlier), was thinking about abandoning physics to go become a historian instead. We now know that the two smudges came from the spinning electron, and not from its going around and round in an orbit.

Different fates for different folks

Phew, think about how many stars have to align for physics to move forward! A cold winter morning, an oath to prove Bohr wrong, a generous Wall Street tycoon, some cheap cigars and an obliging chemical reaction, and a railroad strike!

By the way, Stern never fulfilled his oath and continued to work on physics, till weeks before he was scheduled to be expelled by the Nazis from his university.[40] The Swedes eventually[41] handed him the Nobel prize in 1943, but ironically, not for the Stern-Gerlach experiment, but for measuring the proton's magnetic moment.¶

Stern and Gerlach, whose names are forever tied together in physics textbooks, had vastly different fates later. Stern became a US citizen in 1939 and worked for the allies during World War II. Afterward, he traveled to Europe often but never went back to Germany, and refused the pension that was due him. Gerlach was one of the "good guys" of German physics who refused to join the Nazi party and resisted the crazies who attacked the so-called Jewish physics of Einstein. He rose to lead the German atomic bomb effort, and in an ironic twist of history, was among those German physicists captured by the "electron spinning Goudsmit" and taken to England under house arrest. Goudsmit, who had also become a US citizen, was the scientific leader of the Alsos Mission tasked with locating German physicists to prevent their falling into the hands of the Soviet Union.

Decades later, Goudsmit would rant against historians who fabricated "pretty stories" about scientific discoveries. He wrote, "Luck and random events play a much larger role than people are ready to admit. They (the historians of physics) present things as if the whole of physics was created

---

¶Arguably, this represents yet another of the many goofs by the Swedish Academy. I think that almost all physicists would agree that the Stern-Gerlach experiment is far more important than the measurement of the proton's magnetic moment.

by a handful of geniuses. This is completely unfair to the many physicists whose work enables the great discoveries of the geniuses."[42] I hope that my brief retelling of the discovery of electron spin will persuade the reader to take popular accounts of physics discoveries with a somewhat larger grain of salt.

A quantum scarf factory

How did the Stern-Gerlach experiment inspire my scarf factory analogy? Well, the factory producing a stream of scarves is the analog of the oven emitting a beam of silver atoms. Let us label the three Cartesian axes x, y, and z, and, for ease of writing and talking, call them east-west, north-south, and up-down respectively. Then we could refer to the "heart" of the Stern-Gerlach apparatus with its inhomogeneous magnetic field pointing in the up-down direction as $(SG:z)$. Silver atoms passing through $(SG:z)$ are sorted into those pointing up and those pointing down. See the left half of figure 8(a). The apparatus $(SG:z)$ corresponds to the elf who sorts scarves into red or blue scarves in our fable.

The trucks correspond to what particle experimentalists call beam dumps. An unwanted beam of atoms (molecules or elementary particles) is sent crashing into a block of iron, concrete, or whatever material[43] is on hand, to be absorbed and never heard from again. The right half of figure 8(a) shows the beam of atoms with magnetism pointing up being dumped.

But we could orient the Stern-Gerlach apparatus whichever way we want, at least as a thought experiment. In particular, rotate a $(SG:z)$ into what we call $(SG:x)$, namely a Stern-Gerlach apparatus with its inhomogeneous magnetic field pointing in the east-west direction, corresponding to the elf who sorts scarves into silk or woolen scarves. Thus, sending scarves through two sorting rooms in succession simply corresponds to connecting an $(SG:z)$ to an $(SG:x)$. See figure 8(b).

We could also have an $(SG:y)$, namely a Stern-Gerlach apparatus with its inhomogeneous magnetic field pointing in the north-south direction. We won't need it here, but for the sake of completeness let us say that it corresponds to a room with an elf sorting scarves into those with tassels and those without tassels.

Thus, at the risk of sounding tedious but to make sure that the reader follows, let me describe figure 8(a) and 8(b) again. Two beams of silver atoms emerge from $(SG:z)$ and the spin up atoms are dumped. The remaining

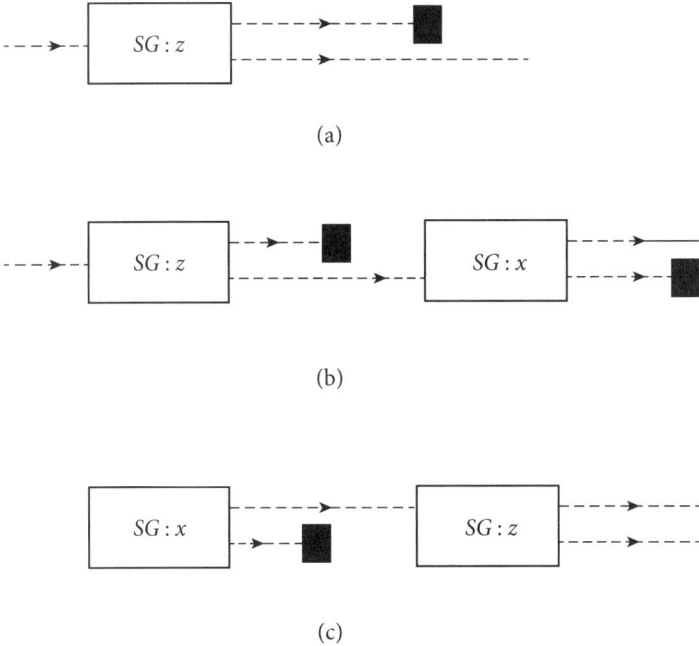

Figure 8. (a) Silver atoms passing through $(SG:z)$ are sorted into those pointing up and those pointing down, with the former dumped. (b) The output of $(SG:z)$ in (a) is fed through a $(SG:x)$. (c) The output of $(SG:x)$ in (b) is fed through a $(SG:z)$. Beam dumps are employed strategically in all three cases. Compare with figure 2.

atoms are all spinning down, and then sent through an $(SG:x)$ and emerge spinning east and west. The beam of atoms spinning west is then dumped.

Regeneration

Now we move on to figure 8(c). The silver atoms spinning east are then sent through another $(SG:z)$, and the quantum surprise is that half of the atoms emerge spinning up, and half spinning down. Physicists say that the spin up atoms that "died," namely were sent to the dump, have regenerated, or, if you prefer to be dramatic, have been "reborn."[44]

By now you could probably figure out the Stern-Gerlach set up corresponding to the set up in figure 3. The two beams in figure 8(b) coming out of an $(SG:x)$ are sent into an $(SG:z)$. Lo and behold, only atoms spinning down emerge!

Physicists say that in the quantum world an atom spinning down is a linear combination or superposition of an atom spinning east and an atom

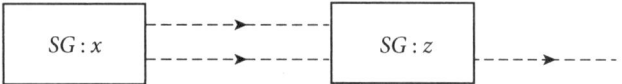

Figure 9. Two beams coming out of an $(SG:x)$ are sent into an $(SG:z)$. Compare with figure 3.

spinning west. This would make no sense in the everyday world. The analogous statement would be to say that a blue scarf is a linear superposition of a silk scarf and a woolen scarf. In a quantum course the linear superposition principle is one of the first items fed to undergraduates.[45]

Not just probabilities, but probability amplitudes: interpretations of quantum mechanics

At this point, I need to go off into an apparent digression, but necessary even for a minimal overview of quantum physics such as offered here. Some readers are aware that quantum physics traffics in probabilities,** but there is a lot more than that. The probabilities in quantum physics are constructed out of complex probability amplitudes.

The very concept of a complex probability amplitude most certainly does not exist in everyday life. I must spend a few moments at least to mention, in barest bone terms, what probability amplitude means. A tiny bit of math is needed here. I could not do better than to quote Feynman: "People who wish to analyze Nature without using mathematics must settle for a reduced understanding."

A probability amplitude is described by a complex number, which could be visualized as an arrow in a plane, that is, as a 2-dimensional vector. Thus, a complex number, traditionally denoted by z, is completely characterized by the length of the arrow and by the angle φ the arrow makes with respect to some reference direction, usually taken to be the x-axis. See figure 10. In contrast to probability amplitude, probability is, and has been since time immemorial, described by a real number between 0 and 1.

In everyday life, if there are several mutually exclusive ways for an event to occur, we add the probabilities associated with each of these ways to obtain the probability for the event to occur. Any Las Vegas gambler[46]

**Probabilities are intrinsic to quantum physics, in sharp contrast to the occurrence of probabilities in gambling, which is simply due to our ignorance of the initial conditions, such as the force with which the roulette wheel was spun.

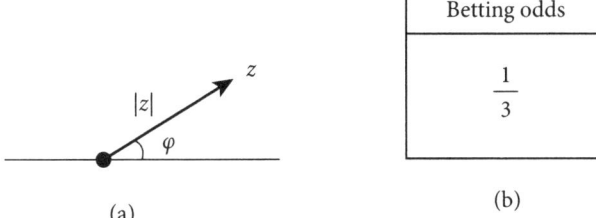

Betting odds
$\dfrac{1}{3}$

(a)   (b)

Figure 10. (a) A probability amplitude in quantum physics. (b) The probability circulating at a race-track that Quantum Wind will beat Classical Gas. This figure underlines the vast distinction between the often confounded concepts of probability amplitude and probability.

could tell us that. For example, the probability of obtaining either a 5 or a 6 in one throw of the dice is evidently $\frac{1}{6} + \frac{1}{6} = \frac{1}{3}$.

In quantum physics, in contrast, we add the probability amplitudes. At the end of the calculation, the probability is given by the absolute square of the sum of the probability amplitudes. (The absolute value of z (denoted by $|z|$) is just the length of the arrow, and the absolute square is the square of the absolute value.) The directions the probability amplitudes are pointing in matters! Many of the mysteries of quantum physics flow from this fact. We will come back to this in chapter 10.

Interpretation of quantum mechanics

When Newtonian mechanics or Maxwell electromagnetism burst onto the stage, physicists did not go into a frenzy about how to "interpret" them. These theories are couched in everyday terms. But from its very beginning quantum mechanics was beset by debates about what it "means." This debate about interpretation has occupied some of the deepest minds well into the 21st century. The subject is extremely confusing and more often than not, very poorly explained in obscure treatises. Learning the nuts and bolts of quantum physics is, for most physics students, much easier than understanding what the warring interpreters are saying. For many years, the debate between people who favor one or the other inter-pretation often lapsed into metaphysics. Hence, most physicists did not pay the slightest attention and simply followed the dictum[47] "Shut up and calculate."

Almost by necessity, courses in quantum physics are taught in the shut up and calculate tradition. Professors do not want to get bogged down by debates about reality that almost border on the metaphysical, and students want to "get on with it" and learn how to calculate.

When I was starting out in my career, an eminent physicist told me to wait until I was old before worrying about the interpretation of quantum physics. Get tenure first, worry later! Over the decades I neglected to ask him how old is old, and then he passed away. Now that I am old I have no idea whether or not I am old enough.

In a sense, the debate over the interpretation of quantum physics means to attach words to a calculation and so is bound to be a confusing muddle, given that words, with all their shady connotations, are derived from every-day life. Unless the debaters carefully define every word such as measure, reality, probability, and so on these discourses too often end up as almost meaningless hot air shrouding a never never land inhabited by philosophers or would-be philosophers.

But there is a remarkable phenomenon that you would not observe in the philosophy department. Two physicists could be arguing vehemently about the interpretation of quantum physics, and yet, if they were told to sit down and calculate what they would predict for a specific quantum process that experimentalists can measure, they would always agree! Physicists could differ on the words attached to the mathematical symbols but yet totally agree on the math. In my more than humble opinion, this ongoing, and some would say never-ending, debate is to a large extent futile, since words necessarily evolved out of everyday life. I truly do not know[48] what it means to say that the universe branches every time an observation is made, as in the so-called many-world interpretation of quantum physics.

The Copenhagen interpretation of quantum physics

The Copenhagen interpretation, proposed by Niels Bohr and others and taught to students as the standard, goes as follows. To make a measurement on a quantum system, we necessarily have to bring it into contact with a measuring device, which is usually macroscopic and therefore treated as a classical system. In quantum physics, an observable[49] is some physical and measurable property, such as the position of a particle or the orientation of the spin of a particle.

Let's say we want to measure some observable we will denote by S. Before the measurement, the quantum system has a probability amplitude c_j, a complex number,[††] of being in a state on which the measurement will yield the value s_j, a real number.[‡‡] Here $j = 1, 2, \cdots, n$ with n the number of possible states.[50] Suppose the measurement gives the value s_7. (I pick 7 just to be definite in what follows; the number 7 has no special meaning.) The probability of obtaining this value is equal to $|c_7|^2$, namely the square of the absolute value of the complex number c_7.

In case the reader wonders where these strange assertions come from, they are postulates whose consequences have been verified countless times since the second decade of the 20th century. You could regard them as experimentally confirmed statements, the totality of which defines quantum physics.

Before the measurement, we are not allowed to say which state the quantum system is in, only the probability amplitude of its being in various states. After the measurement, we can say with certainty[51] that the quantum system is in the state 7. A subsequent measurement (of the same observable S) would show that the system is in state 7 with 100% probability.

During the measurement, according to the Copenhagen interpretation, the system is "magically" thrown into the state 7. The term "collapse of the wave function" is usually taught to students (including poor old me as an undergraduate) but is also objected to by some authors for many reasons. The innocent reader can now see why the debate could be vociferous and never ending: almost every word used here, measure, system, thrown, collapse, during, etc., has been hashed and trashed over. Just to give you a flavor of what physicists argue about, let us ask how the measuring device could be treated as classical if it is part of a quantum universe. Where exactly is the dividing line between the quantum world and the classical world? Some authors would treat the measuring device as quantum also, and say that only the correlation between the two quantum systems which came into contact during the measurement process is meaningful.

To summarize, there is no disagreement about what happens before and after the measurement, only during the measurement.

[††]Here is why I need that apparent digression, telling you about probability amplitudes and their being complex numbers.

[‡‡]Since s_j is a number experimentalists actually measure, it has to be real.

Spooky action at a distance and entanglement

The proverbial guy and gal in the street often associate Einstein exclusively with relativity. But in fact he was also a leading founder of quantum physics.[52] (Such is his greatness!) But as time went on, the grand old man of physics became increasingly disturbed by what he called the spooky quantum, which he denounced[53] in an influential paper published in 1935 with Boris Podolsky and Nathan Rosen.

This paper, known to physicists as EPR,[54] was sharpened in 1951 by David Bohm into its modern and much clearer form, and led him and others to propose that the microscopic realm is actually described by an unknown classical theory with hidden variables that we have no access to, at least not yet, so that the probabilistic nature of quantum mechanics comes from our lack of knowledge of these hidden variables. I will come back to hidden variable theory, but first, let me sketch for you the feature of quantum physics that bothered Einstein and his friends so much.

Bohm imagines a particle X without intrinsic spin just sitting there (hence also without orbital angular momentum), and then suddenly disintegrating into two particles. (The two particles were both taken to carry spin $\frac{1}{2}$, specifically an electron and a positron, and almost all modern textbooks present the theoretical discussion of EPR in terms of two spin $\frac{1}{2}$ particles. This is why I went through a long discussion about the discovery of spin $\frac{1}{2}$. However, almost all experiments utilize a decay process in which an atom emits two photons back to back, because experimental physicists find it much easier to detect and manipulate photons than electrons. In particular, an atom could be coaxed into an excited state, from which it then quantum jumps back down to the lowest energy state emitting two photons moving away in opposite directions. A photon actually carries one unit of angular momentum, rather than half a unit. I hesitate to mention all this because it is not directly relevant to our story here, but one of my colleagues who read the manuscript urged me to do so lest I confuse some of the more knowledgeable readers. The following discussion is written, like most textbooks, about two spin $\frac{1}{2}$ particles moving off in opposite directions, nearly at light speed, rather than two photons. Keep in mind that the experiments are done with photons.)

Crucially, from angular momentum conservation, the two particles must be spinning in opposite directions: if one of them is spinning up, the other must be spinning down. The two particles are said to be

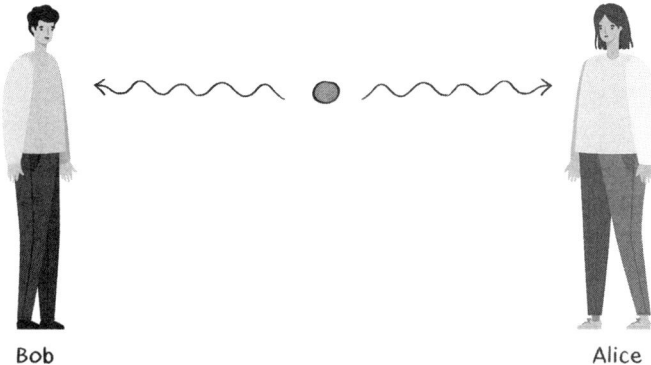

Figure 11. Two photons (in most actual experiments) or two electrons (in most textbooks) spinning in opposite directions and quantum entangled with each other racing toward Alice and Bob.

quantum entangled.[§§] The crux of Einstein's discomfort is that no matter how far apart the two particles are from each other, the laws of quantum physics decree that their spins continue to be entangled.

Meanwhile, two intrepid experimental graduate students, traditionally named Alice and Bob by physicists because of their convenient initials A and B, are stationed light years apart, away from earth in opposite directions, to detect and measure the spin of the two particles speeding toward them at (or near if they are electrons) light speed. Because of momentum conservation, one particle is zinging toward A, the other toward B. See figure 11.

For ease of writing, I will now write the quantum state in which the particle is spinning up as $|\uparrow\rangle$, exactly as how professional quantum physicists do it.[¶¶] Similarly, I write the quantum state in which the particle is spinning down as $|\downarrow\rangle$.

The quantum state in which the particle spinning up is going toward A and the particle spinning down is going toward B is then written as $|\uparrow\rangle_A |\downarrow\rangle_B$. By agreement, physicists soon stop writing the subscripts A and B and lump the two kets into one, thus writing this state as $|\uparrow\downarrow\rangle$. The

[§§]One of the earliest experiments on entanglement was done by C. S. Wu, already mentioned in chapter 2, and her student Irving Shaknov, *Phys. Rev.* 77 (1950), 136. See the discussion in J. J. Sakurai, *Advanced Quantum Mechanics* (1967), page 227.

[¶¶]This so-called bra and ket notation, which I already used earlier, was introduced by Paul Dirac; $|\uparrow\rangle$ is known as a ket. Its counterpart $\langle\uparrow|$, known as a bra, does not enter our discussion here. The peculiar names came from Dirac imagining taking a bracket $<\cdots>$ apart. Isn't theoretical physics fun? It is full of these really dumb nerd jokes.

standard convention that, reading from left to right, we say that the particle in the first position is going toward A, and the particle in the second position is going toward B.

Wait! The attentive reader objects. "How do you know that the spin up particle is going toward A, and not B? There should be a 50-50 probability that it is the other way around."

Yes, you are right! It could also be described by the ket $|\downarrow\uparrow\rangle$. Indeed, the state $|s\rangle$ is actually given by[***]

$$|s\rangle = \frac{1}{\sqrt{2}}\left(|\uparrow\downarrow\rangle - |\downarrow\uparrow\rangle \right)$$

Thus, the probability amplitude that the spin down particle is going toward A and the spin up particle is going toward B equals[55] $-\frac{1}{\sqrt{2}}$. The corresponding probability equals $(-\frac{1}{\sqrt{2}})^2 = \frac{1}{2}$, exactly as you intuited. Similarly, the probability amplitude that the spin up particle is going toward A and the spin down particle is going toward B equals $\frac{1}{\sqrt{2}}$, corresponding to a probability of $\frac{1}{2}$.

The state $|s\rangle$ shown here is known as an entangled state. The spins of the two particles are said to be entangled. In recent years, I am amused to see words like entanglement seeping into the consciousness of the intelligentsia.

Are you bothered? No? Yes? Well, Einstein and his two friends were bothered out of their minds.

Each time Alice measures the spin of the particle coming into her detector, she knows instantly the spin of the particle Bob detects. If she measures up, he must measure down, and vice versa. In other words, in the Copenhagen interpretation, as soon as Alice measures the spin to be up, the state $|s\rangle$ instantly collapses to the state $|\uparrow\downarrow\rangle$.

Suppose the experiment is repeated many times, with Alice and Bob writing down their measurements in a notebook. Later, when they return to earth, they compare notebooks. Alice had written $+ + - + - - - + \cdots$, and Bob $- - + - + + + - \cdots$: there is perfect anticorrelation between their data. But while they were doing the measurements, no communication was possible between them. Einstein called this *spukhafte Fernwirkungen*, "spooky action at a distance."

[***]The minus sign in $|s\rangle$ is dictated by a branch of mathematics called group theory and is not relevant for our discussion here.

Why are physicists spooked?

"You silly physicists," the reader says, "why are you so worked up about this anticorrelation?"

To see why, let us imagine hiring an elf to send out gift boxes to Alice and Bob while they are light years apart, one to each student, with strict instruction to the elf to pack two scarves of opposite color, one in each box. Then as soon as Alice opens her box and sees that her scarf is red, she would instantly know that Bob received a blue scarf. And vice versa.

You ask, "What is so disturbing about this?" Well, thus far, nothing.

But quantum physics is way more subtle than this,[†††] and the attentive reader may have already noticed the key point. "What is so special about up and down? Why are you privileging the z-axis?"

Absolutely right! I could have also written the state $|s\rangle$ as

$$|s\rangle = \frac{1}{\sqrt{2}}\Big(|\rightarrow \leftarrow\rangle - |\leftarrow \rightarrow\rangle \Big)$$

where, as you could have guessed, physicists denote spin pointing east and pointing west as $|\rightarrow\rangle$ and $|\leftarrow\rangle$ respectively. Alice could suddenly decide to measure the spin of the particle along the x-axis. But now, the instant she measures the spin to be, say, pointing east, she knows that the spinning particle in Bob's detector is pointing west. Crucially, now they have lost all knowledge of the whether their particles are pointing up or down.

In my silly scarf analogy, suppose that Alice and Bob were instructed to put on blindfolds before opening their gift boxes, but could use their fingers to determine whether their scarves were silk or woolen. What does not correspond to everyday life is that after doing this, they would lose their abilities[‡‡‡] to see the color of those scarves. (The loss is temporary: physicists are not cruel. The blindfolds on Alice and Bob are removed as soon as the scarves are taken away and stored in vaults to which the two students have no access, so they could wait for the next gift box to arrive.)

[†††] Some ill informed people would stop at the preceding paragraph and proclaim that they understand quantum physics. I have seen accounts describing an eccentric person always wearing a red sock on one foot and a blue sock on the other. The writer then marvels at the fact that if he observes the color of the sock on one foot, he would immediately know the color of the sock on the other foot. Surely the reader recognizes this as a triviality, yet another confused rant on the web.

[‡‡‡] This represents another version of the well-known uncertainty principle as usually stated: if you know the position of a particle exactly, then you lose all information about its momentum, and vice versa.

Indeed, we could continue this line of reasoning and see that the x-axis is no more special than the z-axis. We could measure the spin along any axis, oriented at some arbitrary angles. Then the very same quantum state becomes $\frac{1}{\sqrt{2}}(|\nearrow\swarrow\rangle - |\swarrow\nearrow\rangle)$. In other words, in quantum physics, the very same state could be written identically in many forms:

$$|s\rangle = \frac{1}{\sqrt{2}}\left(|\uparrow\downarrow\rangle - |\downarrow\uparrow\rangle\right) = \frac{1}{\sqrt{2}}\left(|\rightarrow\leftarrow\rangle - |\leftarrow\rightarrow\rangle\right)$$

$$= \frac{1}{\sqrt{2}}\left(|\nearrow\swarrow\rangle - |\swarrow\nearrow\rangle\right)$$

Indeed, since in the last form the two opposing arrows (although drawn at 45°) could be inclined from the vertical at any angle between 0° and 180°, the state $|s\rangle$ could be written in infinitely many apparently different ways.

More generally, Alice and Bob need not measure the spin along the same axes: they are not allowed to—indeed, not able to even if they want to—communicate with each other. They cannot tell each other which axis they feel like using. I am merely keeping things simple here, but of course quantum physics tells us what the corresponding probability amplitude would be.[56]

Hidden variables?

The most celebrated attempt to escape from the spookiness of quantum physics is known generically as theories with hidden variables, first introduced by David Bohm, as I mentioned earlier. The conceit is that at a deeper level, physics is still classical, but contains hidden variables that could be tuned in such a way as to produce the same experimental result as quantum physics. (I don't know about you, but I am already turned off even writing this sentence. It sounds so contrived! No wonder as a student I did not know about this alternative theory and none of my professors ever mentioned it. Looking back, I now realize that the university had sworn to protect the innocent young.)

But then in 1964 the Irish physicist John Bell showed that a hidden variable theory must satisfy a class of inequalities, which quantum theory can evade. Unhappily, Einstein died in 1955, so that we do not know what his reaction to Bell's revelation might have been. It would be tempting to write that Bell stunned the physics world by pointing out that quantum theory and hidden variable theory could be put into a joust experimentally, but in fact Bell's paper was roundly ignored by theoretical physicists.

Fortunately, not by experimental physicists. Three independent series of experiments have since showed conclusively[§§§] that Nature violates Bell's inequalities.[57] Quantum physics triumphs!

Bell's inequalities

Bell's inequalities is actually a statement about classical physics, as I just said. It states that in a classical theory, such as a hidden variable theory, probability correlations must satisfy a set of inequalities.

A common misconception is that it requires a deep mastery of quantum physics to derive Bell's inequalities. Actually, since it is a statement about classical physics, you don't have to know any quantum mechanics. Furthermore, to the extent that it is a statement about probability correlations, you barely have to know much physics. You could derive it by drawing a simple Venn diagram or by making a table of various possibilities.[58] Of course, to show that quantum physics could evade Bell's inequalities does require a knowledge of quantum physics: you could hardly show that something can violate a constraint without knowing anything about that something.

My presentation of Bell's inequalities is adapted from that of Lorenzo Maccone,[59] who in turn based his proof on John Preskill's version of a suggestion by David Mermin. I will continue to use scarves in my explanation now that the reader is already "familiar" with them, but I should emphasize that since we are deriving Bell's inequalities in the classical world, I am no longer talking about those "magical" scarves sorted by elves, but just everyday commonsense garden-variety scarves that we purchased at the local department store. (If you prefer, I could talk about shirts instead.)

Please read the following carefully, and keep in mind that it involves nothing more than high school level probability theory.

To reward Alice and Bob for their arduous work, we want to gift them with a scarf each. Let's suppose that scarves have three attributes, color, fabric, and tassels, to be denoted by C, F, and T respectively. Each attribute has only two values, which we denote by $+$ and $-$ as indicated in this table.

	C	F	T
$+$	blue	silk	tassels
$-$	red	wool	no tassels

§§§The physics Nobel prize for 2022 was given to Alain Aspect, John Clauser, and Anton Zeilinger for their pioneering experimental work.

With this notation, we could write $++-$ to indicate a blue silk scarf without tassels, $+-+$ a blue wool scarf with tassels, $--+$ a red wool scarf with tassels, and so on. With 3 attributes each taking on 2 possible values, we have $2 \times 2 \times 2 = 8$ different kinds of scarves: namely $+++$, $++-$, $+-+$, $-++$, $+--$, $-+-$, $--+$, $---$.

We told an associate to go buy two identical scarves, and to decide which of the 8 kinds to buy according to some preassigned probabilities. We could pick any 8 real numbers between 0 and 1 to assign to the probability $p(+++)$, $p(++-)$, $p(+-+)$, $p(-++)$, $p(+--)$, $p(-+-)$, $p(--+)$, $p(---)$ for each of these 8 different kinds of scarves. (These 8 numbers have to add up to 1 of course.) For instance, we happen to know that Alice and Bob both like blue scarves without tassels. Thus, to please them, we could set $p(++-) = 0.5$ and $p(+--) = 0.47$, with all the other 6 possibilities set to equal 0.005. (The numbers are chosen so that $0.5 + 0.47 + (6 \times 0.005) = 1$, of course.) That they will get blue silk scarves without tassels, that is, $++-$ in the notation used here, is then a hundred times more likely than their getting a red wool scarf with tassels, that is, $(--+)$. These illustrative numbers also serve to underline the fact that there is no funny business from quantum physics here, just the everyday understanding of probability. Not a word about the mysterious probability amplitudes in this discussion.

We give Alice and Bob each a gift box containing a scarf, assuring them that the two scarves are identical. (For this part of the story, they don't have to be separated by light years. Indeed, they could be in the same room, and for that matter, they could even be married by now.)

Each possible scarf has three attributes, color, fabric, and tassels, denoted by C, F, and T respectively, and each one of these attributes could take on two values, $+$ or $-$. Denote the attributes by X and Y, each of which could take on three values, C, F, or T.

Let us now define $P_{\text{same}}(X, Y)$ as the probability that the value of attribute X (be it $+$ or $-$) of Alice's scarf is the same as the the value of attribute Y of Bob's scarf.

As a specific example, $P_{\text{same}}(C, F)$ is the probability that the attribute C of Alice's scarf and the attribute F of Bob's scarf take on the same value. In other words, this is the probability of Alice getting a blue scarf (that is, $C = +$) and Bob getting a silk scarf (that is, $F = +$) or of Alice getting a red scarf (that is, $C = -$) and Bob getting a wool scarf (that is, $F = -$). But since they are getting identical scarves, in the first case they both got a blue silk scarf, with or without tassels, and in the second case they both got a red woolen scarf, with or without tassels. Again, remember that quantum weirdness is not involved here: Bell's inequalities is a statement about the classical world.

At this point, we might think that there are $9 = 3 \times 3$ possible probabilities $P_{\text{same}}(X, Y)$ since X and Y could each take on 3 values. But there are actually only 3, as we will now point out.

Since the two scarves are identical, $P_{\text{same}}(C, C) = P_{\text{same}}(F, F) = P_{\text{same}}(T, T) = 1$. For instance, if Alice got a silk scarf, then so did Bob, and if she got a woolen scarf, then so did he: $P_{\text{same}}(F, F) = 1$. We are merely restating the condition that the two scarves are identical.

It looks like there are 6 other probabilities to worry about. But you could see that $P_{\text{same}}(X, Y) = P_{\text{same}}(Y, X)$. Perhaps it is clearer if I say it in words. Compare two probabilities: the probability $P_{\text{same}}(C, T)$ that the C of Alice's scarf is the same as the T of Bob's scarf, and the probability $P_{\text{same}}(T, C)$ that the T of Alice's scarf is the same as the C of Bob's scarf. But of course the two probabilities are equal because the two scarves, regardless of who now owns them, are identical by assumption; in other words, if we erase the words "Alice's" and "Bob's" in the preceding sentence, the two probabilities we are comparing are tautologically equal!

Indeed, since the story is taking place in the classical world and since Alice's scarf and Bob's scarf are identical, Alice doesn't even need Bob to be around, and vice versa. The reader may have already realized that to determine $P_{\text{same}}(C, F)$ for instance, Alice could have just looked at her scarf to see if it is blue silk or red wool (regardless of whether it has tassels or not) rather than blue wool or red silk. (To obtain a probability, the story would have to be repeated many times.)

We haven't even talked about the quantum experiment; that will come later. To mimic the quantum world, we would have to impose some peculiar constraints. We could say that Alice could glance at the color but she is not allowed to touch the scarf, while Bob could touch the scarf but he is not allowed to look at it. (Some story could be made up as in the story told earlier, the scarf given to Alice is wrapped in heavy translucent plastic, while Bob is given his scarf in total darkness. Alice has dim color vision but no sense of touch, and Bob is the opposite.) Never mind for the moment, let's press on.

So there are only 3 probabilities left to think about: $P_{\text{same}}(C, F)$, $P_{\text{same}}(C, T)$, and $P_{\text{same}}(F, T)$.

At first sight, it would seem that these 3 probabilities $P_{\text{same}}(X, Y)$ for $X \neq Y$ could be freely set by us or by the shopper we sent. For instance, $P_{\text{same}}(C, F) = 0.7$ means that there is a 70% chance that the scarves are made of blue silk or red wool, but only 30% chance that they are made of blue wool or red silk, regardless of whether they have tassels or not. (Look at the table given just now; I am just reading off it.)

The probabilities $P_{\text{same}}(X, Y)$ for $X \neq Y$ are determined by the 8 probabilities $p(+++)$, $p(++-)$, $p(+-+)$, $p(-++)$, $p(+--)$, $p(-+-)$, $p(--+)$, $p(---)$ we or the shopper chose, as will be indicated in an appendix; alternatively, you could write them down now using the everyday understanding of the word "probability." (For instance, $P_{\text{same}}(C, F) = p(+++) + p(++-) + p(--+) + p(---)$.) As noted earlier, these 8 probabilities could be any 8 real numbers between 0 and 1 that add up to 1.

As I said just now, we might have thought that we could freely set the 3 probabilities $P_{\text{same}}(X, Y)$ for $X \neq Y$. But no, says Bell, these 3 probabilities $P_{\text{same}}(X, Y)$ for $X \neq Y$ cannot be freely set by us or our minion. Instead, they must satisfy this inequality:

$$P_{\text{same}}(C, F) + P_{\text{same}}(C, T) + P_{\text{same}}(F, T) \geq 1$$

Perhaps you could sit down and derive it yourself. Yes, you can do it! As you can see and as I said, it was not an exaggeration to say that the derivation of Bell's inequality does not require mastery[60] of quantum physics: it doesn't even require mastery of classical physics. For those who need help, two derivations of this inequality will be given in an appendix.

Counterfactual or prefactual definiteness

Quantum mechanics makes absolutely no sense.[61]

Maccone gave a careful analysis of the two implicit assumptions behind Bell's inequalities: Einstein locality and counterfactual definiteness, which we will now define.

First, Einstein locality means that information could not propagate faster than the speed of light. In our example, this means that the act of Alice opening her gift box cannot somehow affect or change the content of Bob's box.

Second, Maccone defines a physical theory to be "counterfactual definite" if measurements in this theory uncover pre-existing properties. Thus, classical physics is a counterfactual definite theory. In this kind of theory, it is meaningful to assign properties to an object independent of whether or not measurements to determine these properties will be made. Again, in our story set in the classical world, before Alice opens her box, she (and we) have no idea whether the scarf inside is red or blue. Yet everybody present firmly believes that the redness or blueness of the scarf is already fixed and is not affected by her opening the box. The color cannot change magically. In other words, color is real (if you want to use that term), and part of the reality which commonsense tells us exists.

A weirdo nerd friend could, for some unfathomable reason, refuse to reveal to us the color of a scarf he owns, but instead asks us to guess, telling us that there is a 70% chance that it is blue, and 30% chance that it is red. But for sure his scarf has a definite color even though we have never seen it. Since he knows the color of his scarf, this use, or rather abuse, of the word "probability" in everyday life is merely to tease us into guessing the color, along the line of "Guess what I did this weekend!" That color exists independent of the observer is what we might call real. No one (besides a true crazy) would say in daily life that the color of our friend's scarf is a meaningless construct until we actually see the scarf. Whatever our weirdo friend actually did on the weekend could not change by the time he asked us, on Monday, to guess what he did.

When I first came across the term "counterfactual definiteness," I recoiled as if this is yet another of those highfalutin' complicated sounding terms way above my pay grade, suitable only for pipe puffing philosophy professors having elaborate lunches at the faculty club[62] while physicists munch sandwiches at lunch seminars. But now I like this term, and certainly prefer it to wishy-washy terms such as elements of reality. Before Alice opens her box, the color of her scarf has not yet been factually established but yet it exists as part of everyday reality. (I might have preferred "prefactual definiteness.")

	counterfactual or prefactual definite
classical physics	yes
quantum physics	no

Maccone said that he preferred to avoid philosophically laden words like real, reality, and so on and so forth, which may mean different things to different people. I applaud his use of the unusual term "counterfactual definiteness." Indeed, many years ago I read a description of Bell's inequalities and I circled the term "elements of reality" the author used. I wrote in the margin, "Whatever that means?"

Since Einstein locality has not been questioned (at least not seriously), we conclude that the experimental violation of Bell's inequalities shows conclusively that quantum physics is not counterfactual definite. Until the quantum state collapses, we have only probability amplitudes, as per the Copenhagen interpretation.

Perhaps it is worth emphasizing again that deriving Bell's inequality is "easy": it is a property of classical physics. It is showing[63] that quantum physics violates the inequality that requires some knowledge of quantum physics, obviously. Having said this, I should at least sketch for you how the experiment[¶¶¶] is done. Let Alice and Bob be stationed far apart from each other. As I said earlier, experimentalists find it easier to send out two photons, rather than two spin $\frac{1}{2}$ electrons.

For narrative convenience, let the two photons be spinning in the same direction (rather than in opposite direction as was described earlier): these correspond to the two identical scarves given one each to Alice and Bob in the classical story. The three attributes C, F, and T correspond to three different axes along which Alice and Bob could measure the spin of the photon they receive. For example, C could be the z-axis, and F and T could be two other axes, each at some angle from the z-axis and from some angle from each other.

Alice and Bob could each choose to measure the spin of the photon they receive along any one of these axes, and they are not allowed, or unable, to communicate to each other which axis they have each chosen. After each measurement, they write down the axis they used, either C, F, and T, and whether they measured $+$ or $-$. Eventually, they meet and compare notes, figuring out the probability they recorded $++$ or $--$, for various choices of the measurement axes, thus obtaining the three numbers $P_{\text{same}}(C, F)$, $P_{\text{same}}(C, T)$, and $P_{\text{same}}(F, T)$.

[¶¶¶] What I am going to tell you presently is not the actual experiment, but in the spirit of the actual experiment, and in conformity with the story just told as much as possible.

A fairly simple calculation using quantum physics then reveals that by the appropriate choices of these angles we could have the sum $P_{\text{same}}(C, F) + P_{\text{same}}(C, T) + P_{\text{same}}(F, T)$ to be less than 1, thus contradicting Bell's inequality. The Nobel celebrated experimenters then set out to measure this sum, finding that the sum is indeed less than 1, thus ruling out hidden variables.

From the spinning electron to our revised notion of reality

So, in journeying through this rather long chapter, we see that the discovery of the spinning electron has led us to question the nature of reality as understood by "normal" everyday humans. Einstein famously said that God does not play dice.[64] But not only does God play dice, he throws the dice where we are not allowed to see them.[65]

The discovery that our world is but an approximate manifestation of an underlying quantum world is surely the greatest discovery in the history of the human race. It is so unexpected, so astonishing in scope, and so far reaching in its implications, our very notion of what we mean by reality has to be revised.

I end this admittedly confusing chapter with an unsolicited advice in a concluding endnote.[66]

I started this chapter with a quote about quantum mechanics, and now I close with another expressing the same sentiment.

> If you are not completely confused by quantum mechanics, you do not understand it.
>
> —John Wheeler

I hardly carried the weight of Bohr, Penrose, and Wheeler, who all assured us that quantum physics made no sense. Well, even as a lowly undergraduate I could have told the professor who fed quantum physics to me and to my friends that what he was saying made no sense. But I was not alarmed either. The classical world is but an emergent approximation to the quantum world.

A note to the young thinking of going into physics whom I envisaged in the prologue. What is the most important thing you should have learned from this chapter? No, not quantum spin; not Bell's inequality either. Those have been done. The historical circumstances around the discovery of electron spin are fascinating, but erudition is not as important in physics as in the humanities.

The most important lesson is not to let an authority figure crush your ideas. (Some might say it is to stay in bed on a cold winter morning, but that's only if you could think of an epoch changing experiment or calculation you might do.)

Appendix: Derivation of Bell's inequality

The proof of Bell's inequality is surprisingly straightforward, considering the mystique that shrouds it. Simply write down the three probabilities that appear in the inequality. First,

$$P_{\text{same}}(C,F) = p(+++) + p(++-) + p(--+) + p(---)$$

Similarly,

$$P_{\text{same}}(C,T) = p(+++) + p(+-+) + p(-+-) + p(--)$$

and

$$P_{\text{same}}(F,T) = p(+++) + p(-++) + p(+--) + p(---)$$

Recalling that by definition

$$p(+++) + p(---) + p(+-+) + p(-+-) + p(-++)$$
$$+ p(+--) + p(-++) + p(+--) = 1$$

we add to obtain the desired inequality

$$P_{\text{same}}(C,F) + P_{\text{same}}(C,T) + P_{\text{same}}(F,T)$$
$$= 1 + 2(p(+++) + p(---)) \geq 1$$

Simply, in words, in the sum $P_{\text{same}}(C,F) + P_{\text{same}}(C,T) + P_{\text{same}}(F,T)$, the probabilities $p(+++)$ and $p(---)$ are each counted 3 times while the other probabilities are each counted once.

That's it! Did you expect something more involved?

A graphical representation using a Venn diagram may be illuminating. See figure 12. The large circle enclosing the two smaller circles represents all possibilities and thus corresponds to probability 1. The area of the circle with vertical lines represents $P_{\text{same}}(C,F)$. The area of the circle with horizontal lines represents $P_{\text{same}}(C,T)$. Thus, in the area outside these two circles $C \neq F$ and $C \neq T$. Since C, F, and T could each take on only two

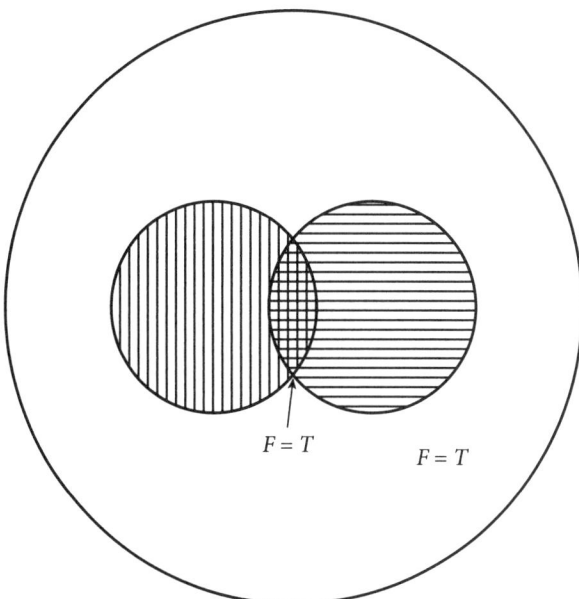

Figure 12. Vertical lines, horizontal lines, three circles, a "pointy oval" shaped region, and Bell's inequality. A detailed explanation is given in the text.

values $+$ and $-$, in this area $F = T$. But in the "pointy oval" shaped region where the circle with vertical lines and the circle with horizontal lines overlap, $C = F$ and $C = T$, and thus $F = T$, there also. So the area where $F = T$ is actually larger than the area outside these two circles $C \neq F$ and $C \neq T$. Hence the sum of the three areas where $C = F$, where $C = T$, and where $F = T$ is greater than the area of the large circle encompassing all. Bell's inequality follows. To me this is a more cumbersome way of expressing what is expressed by the equations presented earlier, but there are some who prefer a pictorial explanation.

Notes

[1] Niels Bohr. In conversation with Heisenberg and Pauli in Copenhagen (1952).

[2] Through a straightforward approximation; by "straightforward" I mean that almost all competent physics and engineering majors, let alone applied math majors, know this method, referred to variously as the stationary phase or steepest descent approximation. See chapter 10.

[3] Unfortunately often by people who don't know what they are talking about.

[4] A better argument invokes the quantization of angular momentum, since unlike energy, the amount of angular momentum cannot go negative.

[5] For those readers who wish to learn more about quantum physics, I recommend chapter 26 of R. Freedman et al., *College Physics*. This is a calculus-free textbook designed for undergraduates planning to go to medical school.

[6] For instance, the first two rows of the periodic table. See *FbN* chapter III.4.

[7] See Brandon Brown's biography of Planck.

[8] Some pedants would probably insist that this is actually the reduced Planck's constant and that the frequency I speak of here is actually the circular frequency. Believe me, I know all that but I do not want to confuse the readers unnecessarily.

[9] See, for example, *FbN*, page 22.

[10] See *FbN*, page 23.

[11] Some of these are difficult for physicists nowadays to appreciate. For instance, we take for granted that in atoms the electrons were electrically attracted to the nucleus. But how did the pioneers know there weren't some other forces? Bohr, for example, had proposed a new force he called the "unmechanische Zwang." What a relief that the Zwang turned out to be fictitious!

[12] S. Goudsmit, *Physics Today*, June 1976; G. Uhlenbeck, *Physics Today*, June 1976; A. Pais, *Physics Today*, December 1989.

[13] Early in my career I had the honor of meeting both of them. Goudsmit was for many years the editor in chief of the American physics journal *Physical Review*, and launched the once-important journal *Physical Review Letters*.

[14] They received their doctorates in July 1927, while their famous paper was published in February 1926. An amusing footnote to history is that Goudsmit had not passed his mechanics exam, and so was forbidden by law to teach high school physics in the Netherlands, but yet taught a graduate course at the University of Michigan in the United States. Unfortunately, American bureaucrats were merely a few decades late in rearing their heads. Now it is no longer a matter of what you know, but what degrees you have.

[15] Receiving his PhD in 1925 and becoming an assistant professor in 1927, he was inspired by Pauli's lecture in Tübingen in 1925.

[16] In our degenerate age, asking a senior person for advice could be a lot more dangerous than getting discouraged; as soon as the young person's back is turned, the senior person might be publishing the idea under his or her own name.

[17] Known as the classical electron radius. See *FbN*, pages 74–75.

[18] Consider the expression for angular momentum $L = mvr$ given in the text for an object of mass m and radius r rotating with velocity v. For a given value of L, if r is tiny, then v has to be enormous. You are invited to plug in the numbers to prove Pauli's point. The trouble, for poor Kronig, is of course that this formula of classical mechanics, known already to Newton, is simply not valid in the quantum world. Do not picture the electron as a tiny billiard ball spinning in space. Rather, in quantum physics, we consider how its probability amplitude transforms under a rotation.

[19] I quote Heisenberg from an interview he gave in 1962: "Perhaps the main mistake of Kronig was that he didn't really fight. I mean, if you have a good idea, then you must not only pronounce it, you must fight for it, because your ideas are never liked by the other people. That's a normal thing." To the young persons I alluded to in the prologue of this book, take this to heart!

[20] Behind his affable façade Ehrenfest actually led an anguished life. He was increasingly depressive, involved in an extramarital affair, and then, at the age of fifty-three, he killed his fifteen-year-old son, who was born with Down syndrome and lived at the Institute for Afflicted Children in Amsterdam, just before killing himself. He was forty-five in 1925, the year of the electron spin.

[21] These two stories suggest to me that both Pauli and Lorentz had already considered and rejected the notion of a spinning electron.

[22] In our topsy turvy age, a young person's mistake may be career ending, while a famous person's error might be laughed off.

[23] This entire story was told by Bohr to Pais in 1946.

[24] "Eine neue Kopenhagener Irrlehre!"

[25] An understanding of what red and blue mean is not relevant to the story told here, but curious readers not residing in the United States might want to consult the map classifying the fifty states as blue or red, shown in https://en.wikipedia.org/wiki/Red_states_and_blue_states.

[26] A. Peres, *Am. J. Phys.* 46 (1978), page 745.

[27] Schrödinger's cat, now a meme and part of our internet culture, only serves to confuse many people, in my humble opinion. My favorite play about physics opens with Schrödinger sitting in an armchair thinking about physics. Suddenly a cat jumps onto his lap. Annoyed, Schrödinger grabs the cat by the neck and flings it out the window. A moment later, Schrödinger's wife walks in and asks her husband if he has seen her cat. At least that is how the playwright imagined the source of Schrödinger's inspiration.

[28] Books that start with the Stern-Gerlach experiment include ones by D. H. McIntyre, by J. J. Sakurai (later with S. F. Tuan and with J. Napolitano), and by R. P. Feynman. The textbook I had as an undergraduate discusses the theoretical framework of the Stern-Gerlach experiment only after more than 500 pages! Yikes. Pedagogy in physics has on average changed for the better, in my opinion.

[29] Thus, the year 2022, when I wrote this chapter, was the centenary of this famous experiment. I subsequently learned that my humanist colleagues were all excited because 2022 was the centenary of the poem *The Waste Land* by T. S. Eliot, which I also studied as an undergraduate alongside the Stern-Gerlach experiment.

[30] For more details about this oath, see *FbN*, page 24.

[31] This reminds me of the fact that most Americans are no longer aware of where the "down" in downtown refers to. Yet the term downtown is used even in cities shaped more or less like a circular blob.

[32] For those readers familiar with the scalar or dot product between the two vectors this energy has the form $E = -\vec{\mu} \cdot \vec{B} = -\mu B \cos \theta$.

[33] The magnetic moment of a charged particle is inversely proportional to its mass. Since the proton is 2,000 times more massive than the electron the contribution of the protons to the magnetic moment of the silver atom is completely negligible.

[34] I copy the list in B. Friedrich and D. Herschbach, *Physics Today*, December 2003.

[35] I keep saying smudges, but due to how the inhomogeneous field was configured in the original experiment these dotlike smudges were actually smudgy lines.

[36] I had long thought of this Goldman as the founder, but the chronology indicates that he was a descendant of the founder.

[37] I am grateful to the late Sandip Pakvasa for first telling me the cigar story. See B. Friedrich and D. Herschbach, *Physics Today*, December 2003, and also S. Pakvasa, arXiv:1805.09412.

[38] Incidentally, in 2003, Dudley Herschbach, a Nobel laureate in chemistry at Harvard, and his collaborators tried to verify the cigar story by blowing cheap cigar smoke at a silver coated glass slide. See the paper cited in B. Friedrich and D. Herschbach, *Physics Today*, December 2003.

[39] By sheer coincidence (?), just as I wrote this, a email arrived from the physics departmental chair warning the faculty that some forms that had to be filled out periodically had changed because some administrator had introduced a new regulation.

[40] See I. Estermann and S. N. Foner, *Am. J. Phys* 43 (1975), 661.

[41] He had the dubious distinction of being the second most nominated person for a Nobel prize with eighty-two nominations, second only to Arnold Sommerfeld with eighty-four nominations.

[42] R. Golub and S. Lamoreaux, page 4.

[43] Particle experimentalists have used a pile of rocks and even a junked battleship, as we will also see in chapter 7.

[44] Regeneration is now observed routinely in some particle experiments. Perhaps it is worth emphasizing that "resurrection" is not involved here. The spin up atoms who went to the dump were absorbed. These spin up atoms are not the "same" as those.

[45] In everyday life, we are most familiar with the linear superposition of waves, water waves, sound waves, and so on. In acoustics, "dead spots" could exist in a poorly designed auditorium due to destructive interference.

[46] See the acknowledgments on page xvi of *QFT ASAP*.

[47] Often attributed wrongly to Feynman but was in fact first enunciated by David Mermin, who has written extensively about the meaning of quantum physics.

[48] I once published a paper with Steve Hsu on the derivation of the Born rule in quantum physics and I had to state in the opening paragraph that I do not share my coauthor's views on the interpretation of quantum mechanics.

[49] There are strict rules regarding what is observable and what is not observable.

[50] For the benefit of some readers, let me say all this in mathematical terms, in the form fed to unsuspecting undergraduates. Before the measurement, the quantum system is in the state $|\psi\rangle = \Sigma_{j=1}^n c_j |s_j\rangle$. In quantum physics observables are represented by operators. The operator S acting on the state $|s_j\rangle$ returns that state multiplied by the real number s_j, viz $S |s_j\rangle = s_j |s_j\rangle$. The probability of the measurement giving the value s_k is equal to $|c_k|^2$. After a measurement giving the value s_k, the quantum system is in the state $|\psi'\rangle = |s_k\rangle$.

[51] By the way, this provides another reason why some authors object to the term "the uncertainty principle," which is however completely entrenched by now.

[52] See, for example, A. Douglas Stone, *Einstein and the Quantum: The Quest of the Valiant Swabian*, 2013.

[53] In an exchange of letters with Schrödinger in 1935, Einstein used the German idiom "ist mir Wurst," literally "it is sausage to me" and meaning "I don't care."

[54] See chapter 15 in R. Golub and S. Lamoreaux for a historical account of the EPR paper.

[55] The mathematically sophisticated reader might wonder why the probability amplitudes here, namely $\pm\frac{1}{\sqrt{2}}$, are both real numbers, while I expended a lot of verbiage about complex numbers. In general, probability amplitudes are complex. The examples given here are chosen for maximal simplicity.

[56] For more mathematically oriented readers: Denote by θ the angle between the two axes Alice and Bob use respectively to measure spin along. Then, if Alice measures spin up, the probability that Bob would also measure spin up equals $\sin^2 \frac{\theta}{2}$. This equals 0 when $\theta = 0°$ and 1 when $\theta = 180°$, as expected. When their axes point in opposite directions (that is, when $\theta = 180°$) if Alice measures spin up, Bob always measures spin up.

[57] Bell's inequalities have subsequently been sharpened by numerous physicists, including Bell himself. Many experiments were actually based on the so-called CHSH inequality proposed by J. F. Clauser, M. A. Horne, A. Shimony, and R. A. Holt, *Phys. Rev. Lett.*, 23 (15), (1969), 80–84.

[58] According to M. Le Bellac, this type of inequality was already known in the 19th century to George Boole, the founder of Boolean algebra and a pioneer of computer science. See Le Bellac, *The Quantum World*, page 49.

[59] Lorenzo Maccone, "A simple proof of Bell's inequalities," *Am. J. Physics*, 81 (2013), 854, and the references therein.

[60] As I said earlier, to show that quantum mechanics violates Bell's inequalities requires knowing quantum mechanics, of course, but actually not that much. Readers who are interested could read L. Maccone, ibid.; McIntyre, chapter 4; Sakurai and Napolitano, pages 238–245.

[61] Roger Penrose, *Quantum Concepts in Space and Time* (Oxford University Press, 1986).

[62] I could say this because I actually had a long lunch with a philosophy professor at the faculty club a few weeks prior to writing this chapter.

[63] Which I evidently cannot do here. See the following endnote.

[64] To which Bohr retorted, "Einstein, stop telling God what not to do."

[65] Attributed to Hawking, in reference to black holes.

[66] Here is an unsolicited advice you could take or leave. You can read popular books on quantum mechanics till you are blue in the face, but yet, if you are able to read a single decent textbook on quantum mechanics, a lot would become clear. The response of people to this advice is typically that they cannot handle the math. What math you need is actually minimal compared to that required by other subjects in physics (such as Einstein gravity). A knowledge of calculus and a nodding acquaintance with matrices and partial differential equations suffice. In fact, you could obtain a basic understanding of quantum physics without knowing all that much about partial differential equations, since they often reduce to ordinary differential equations in simple cases.

4
QUANTUM FIELDS FOREVER: EINSTEIN'S TOTAL LOVE

Dancing fields tell each other how to dance

The universe is pervaded with quantum fields. They are all around us, all the time, murmuring in silence. They intertwine with each other, unceasingly telling each other how to dance.

It took the human race an exceedingly long time to realize this basic truth about the universe. Physics as taught to students is ostensibly concerned with particles, particles such as electrons and quarks, careening about like so many teeny billiard balls. But these particles are now understood to be excitations in the quantum fields.

John Wheeler, my first mentor[1] in physics, is often quoted summarizing Einstein gravity thus: "Spacetime tells matter how to move; matter tells spacetime how to curve." While this beautiful saying is undeniably pithy, the picture of mutual influence is not exactly a novelty with physicists. It was already central to classical electromagnetism, which may be summarized as follows: "The electromagnetic field tells the charged particles how to move; the charged particles tell the electromagnetic field how to vary in space and in time."

With further development in quantum field theory, theoretical physicists realized that matter, comprised of particles charged or uncharged, is actually produced by fields. (See later in this chapter.) Fields are more basic than particles. Thus, a more modern summary of all of physics, not just Einstein gravity, would be

> The universe results from a bunch of quantum fields telling each other how to vary in spacetime.

Dancing fields tell each other how to dance. Quantum field theory in one sentence, in a nutshell so to speak.

We will see in chapter 7 that different types of gauge fields[2] drive the four fundamental interactions in the universe. The gravitational field, which manifests itself as curved spacetime, is also a gauge field. The universe is all fields all the time.

The division between matter and spacetime in the classical context of Einstein gravity in Wheeler's quote is thus somewhat outdated. If we want to separate the universe into two sectors, a better scheme would be to separate the fields into bose fields and fermi fields,* instead of fields and particles.

It is manifestly impossible to treat in a short chapter such a fabulously rich subject as quantum field theory, surely the most intricate of all the verified subjects in physics. I aspire to give you only a flavor. This introduction to quantum fields is intentionally mysterious and hopefully enticing.[†] Please do not expect every sentence to be clearly explained as in a textbook. In the rest of this chapter, I will tell you the historical origin of the concept of fields in classical physics, and how fields become quantized with the coming of quantum physics, leading eventually to the crucial realization that fields beget particles. I will trace the birth of quantum field theory in the early 1930s, its rebirth after being interrupted by World War II, and its present day use as an effective field theory.

Action at a distance

Our common everyday understanding of force involves contact: we can exert a force on an object only if we are in contact with it. In a contact sport such as American football, a linebacker could hardly exert anything on the ball carrier without tackling him. Friendly persuasion is known not to work. And in the movies, a slap is not a slap until the leading lady's palm makes contact with the leading cad's cheek. At the supermarket you can push the shopping cart only if you grip the handle. If you could just hold out your hands and command the shopping cart to move, a crowd would gather and honor you as a wizard.

---

*Bose fields love to party while fermi fields prefer to stay by themselves. See *QFT ASAP*, part VI. And in the yet-to-be-established superstring theory all these fields are unified as different degrees of freedom of the superstring.

[†]Readers so enticed might want to try my *QFT ASAP*. Those with some mathematical background will find a wealth of textbooks available, among which my *QFT Nut*.

Just about the only commonplace example of a force acting without contact is the refrigerator magnet: You can feel the refrigerator pulling on the magnet before the magnet makes contact with the refrigerator. That magnets could act on each other while separated by empty space is most alluring to children, and to physicists as well.

Everyday forces, except for gravity, are short ranged, indeed zero ranged on the length scales of common experience.[3] The palm molecules have to be practically on top of the cheek molecules before the latter could acquire any carnal knowledge of the former.

Gravity is the glaring exception. When the earth pulls Newton's apple down, no hand comes out of the earth grabbing the apple as in a horror movie. Gravity is invisible, thus all the more horrifying to those of us who age and sag.

In the days of old, wise men found it necessary to affix stars and planets to celestial spheres, made presumably of some celestial substance with magical properties, slowly turning around and around.[4] This mechanistic picture would have sounded rather convincing to the ancients. Within this worldview, Newton's proposal that the earth's gravity can pull not only the apple down, but that its invisible arm could reach out across the unfathomable vastness of space and tug at the moon, was manifestly absurd.

Lacking in "faculty of thinking"?

In physics textbooks, students learn about the Newtonian concept of action at a distance. The moon is attracted to the earth; no contact is necessary. More advanced books then point out to the bewildered students that action at a distance is kind of bizarre, and set up poor Newton as a straw man to be attacked.

Very unfair! Newton did fret much about action at a distance. In a 1693 letter to his friend Richard Bentley, he opined:

> That gravity should be innate, inherent and essential to matter so that one body may act upon another at a distance through a vacuum without the mediation of anything else by and through which their action or force may be conveyed from one to another is to me so great an absurdity that I believe no man who has in philosophical matters any competent faculty of thinking can ever fall into it.

Please hold Newton's "anything else" in your thoughts.

Einstein brought the gift of time to Newtonian gravity

Another unsettling feature of Newtonian gravity is that time does not play a role. The attractive force exerted by the earth on the moon is given by the product of the masses of the earth and of the moon multiplied by Newton's gravitational constant G and divided by the square of the distance between them. That's that. Any change in the position of the earth is instantaneously communicated to the moon. In Newtonian gravity, the moon is inexorably yoked to the earth. In turn, the earth is yoked to the sun, and the entire galaxy moves as a collective entity.

How could a moon know instantly that its planet has moved? In the *Principia*, Newton left[5] this conundrum "to the consideration of the reader."

The "reader" who took it up was Albert Einstein. He brought the gift of time to Newtonian gravity. The gravitational field, instead of sitting there like a dope, could henceforth move around. Endowed with dynamics by Einstein, it could even wave and travel through spacetime from eons past to some gravity wave detector on some obscure planet. While Newton had the insight of gravity extending far beyond the moon, as we saw in chapter 2, Einstein allowed the entire universe to evolve and curve under the influence of the gravitational field.

Greatness and audacity: enter the field

> For us, who took in Faraday's ideas so to speak with our mother's milk,
> it is hard to appreciate their greatness and audacity.[6]

Before telling you about this great and audacious[7] concept of field, I should tell you a bit about Michael Faraday. The notion of the field was born out of poverty. Faraday, the preeminent experimental physicist of the 19th century, grew up in Dickensian poverty. While an apprentice to a bookbinder, he came across an article about electricity and magnetism in the *Encyclopedia Britannica*. He later lamented that due to his lack of education he was unable to understand the mathematics of the continental philosophers, by whom he meant the leading French physicists of his time, such as Coulomb and Ampère, who used partial differential equations. Just to give the reader some perspective, undergraduate physics majors these days have to be acquainted with partial differential equations before they can tackle subjects such as electromagnetism and quantum mechanics.[8]

While Faraday's genius manifested itself in the laboratory, he also introduced into theoretical physics the important and fruitful concept of a "field of force," or "field" for short, which he invented in order to visualize what the "continental philosophers" were yakking about.

May the field of force be with you

But what is this field of force postulated by Faraday that Einstein considered to be so "great and audacious," and now known to every child who has seen films on interstellar warfare? As I mentioned, in our everyday experiences, we tend to think of a force being exerted only when contact is made between material bodies. Newton's notion of action at a distance had deeply troubled many thinkers, and now, in the 19th century, electromagnetism demonstrated this ever more dramatically. Coulomb's law, for instance, states that the repulsive force between two like charges is given by the product of the charges divided by the square of the distance between them. Very much like Newton's law of gravity, and also relying on action at a distance.

Like many of his predecessors and contemporaries, Faraday grappled with this philosophical problem. He visualized what was going on by sprinkling iron filings on a piece of paper next to a wire. When a current was turned on, the iron filings would obediently form a pattern. Another pattern was formed when the filings were brought close to a magnet. Eventually, Faraday proposed that a magnet or an electric current produced what became known as a magnetic field,[9] which exerted a force on the iron filings.

Similarly, an electric charge produces around it an electric field of force. When another charge is introduced into this electric field, the field acts on this charge, exerting on it a force in accordance with Coulomb's law.

To say that the earth exerts a force on the moon is an old fashioned, but very convenient, way to talk about gravity. A more modern way is to say that the earth generates a gravitational field, which in turn acts on the moon.

The field is a separate entity

In effect, Faraday introduced an intermediary, the "anything else" in Newton's 1683 letter: Two charges do not act "directly" on each other but they each produce an electric field that acts on the other charge.

Faraday's notion does not explain Coulomb's law; rather, it appears to be merely another way of describing Coulomb's law. A pragmatic physicist at the time might be inclined to dismiss the field as just palaver that did not advance our knowledge one whit.

But this view misses the point. The real content of Faraday's picture, as it turns out, lies in the fact that the electromagnetic field not only can be thought of as a separate entity, it is a separate physical entity. Physicists were to learn later that it makes perfect physical sense to talk of the energy density, for example, contained in an electromagnetic field.

So, the electromagnetic field exists as a physical entity, separate from the charges and currents that produce it, regardless of how it is produced[10] and of whether or not another charge is introduced to feel the effect of the field.

Even more amazingly, the electromagnetic field, being a separate entity, could bid farewell to the charges and currents that begat it, leave home, and take off on its own, humming its way through spacetime.[11]

By the time Maxwell burst onto the scene, a century or so of arduous experimental work had already been distilled into various laws. For instance, one equation states how a magnetic field varying in time produces an electric field varying in space. This expresses Faraday's discovery that by moving a magnet around a wire, that is, by varying the magnetic field near the wire, he could produce an electric field that pushed charges forward in the wire, thus generating a current.

"Was it a God who wrote these signs?"

Maxwell codified these laws into mathematical statements, known ever since as Maxwell's equations. Much to his surprise, this time honored collection of equations, many already written down by "continental philosophers," was not mutually consistent. Remarkably, he discovered that by adding a term to one of the equations, he could bring all of them into harmony.

This additional term turned out to be crucial for our modern life that the uninformed take so lightly and for granted. Armed with the correct equations, Maxwell made a truly epochal discovery: the existence of electromagnetic waves. Roughly speaking, if we have in a region of space an electric field changing in time, then a magnetic field is produced in the neighboring space. The production of this magnetic field means that it is also changing in time, and thus generates an electric field. Like a ripple on a pond spreading from a dropped pebble, an electromagnetic field propagates out in a wave, undulating between electric and magnetic energy.

For the first time in the history of the human race, we knew what light is! The electromagnetic age dawns almost as an afterthought.

Three stories about three eminent physicists' appreciation of Maxwell's equations before we move on.

Ludwig Boltzmann, another 19th century great[12] whom we will meet in chapter 9, once pointed to Maxwell's equations during a lecture and exclaimed, "Was it a God who wrote these signs?" I and many physics students feel the same way.

Oliver Heaviside,[13] a giant of electromagnetism who rewrote Maxwell's equations into the form now taught to students, recalled in his old age that when he first encountered Maxwell's work, "I saw that it was great, greater, and greatest!"

When Albert Einstein was introduced at Cambridge University as "the man who stood on Newton's shoulders," he replied, "No, I didn't; I stood on Maxwell's shoulders."

What exactly is waving?

A question that a thinking student might ask: What exactly is waving?

The field is waving.

What is a field?

I already told you.

You cannot use the common sense you cultivated from living in the everyday world to understand quantum mechanics, as we saw in chapter 3. In the same way, since you do not live in the relativistic world, you cannot use common sense to insist that something you can hold in your hand is waving.

For water waves, we can see the water waving, but even in this everyday example, the motion of the wave is not the same as the motion of the water. The best analogy I could come up with in my class is the stadium wave at sporting events, zooming around the stadium formed out of human beings standing up and sitting down, but the human beings are not zooming around the stadium, and in any case, they can't zoom that fast. In an electromagnetic wave, the electric and the magnetic field "change" into each other, transporting energy forward, but they are not themselves literally moving forward.

Perhaps, as in a Zen koan, it is the mind of the theoretical physicist that is waving? (I'm kidding.[14])

As real as a rhino

So, we are literally swimming in a sea of electromagnetic fields.

Dear reader, these are not mere words. The crucial, and meaningful statement, is that the field as a physical entity is entirely real. As real as a rhino, saith the Indian American physicist Anupam Garg.[15] And so on the back cover of his textbook on electromagnetism I blurbed that quantum fields are as real as quantum rhinos.

Electron and photon treated unequally

In introducing quantum mechanics in chapter 3, I mentioned that when an electron in an atom jumps from a higher energy level to a lower energy level, it emits a photon, the quantum of electromagnetism. In the reverse process, the electron absorbs a photon and jumps from a lower energy level to a higher energy level.

Incidentally, in the jargon, the emitted photon is said to be created by the downward quantum jump, while the absorbed photon is said to be annihilated by the upward quantum jump. Jargon is not important except that is how theoretical physicists talk, and I should follow that.

This also would be a good place to clarify the terms nonrelativistic and relativistic. We will discuss Einstein's special relativity in chapter 6, according to which nothing can travel faster than the speed of light c. Particles moving at speeds significantly less than c are referred to as nonrelativistic, and physics that applies only to nonrelativistic particles is said to be nonrelativistic. When first formulated, quantum mechanics was decidedly nonrelativistic, which fortunately was adequate for treating atoms smaller than, say, lead.[16] For instance, in the hydrogen atom, the lone electron whirls around at a speed less than one hundredth of c, a real slowpoke in the subatomic world.

Remarkably, in 1928, a mere two years after the Schrödinger equation, Paul Dirac wrote down in a flash of insight his eponymous equation for a relativistic electron. As used in physics, relativistic is an inclusive term: the Dirac equation, while capable of describing a relativistic electron, can also describe a nonrelativistic electron. Indeed, for a slow moving electron, the Dirac equation must, and does, reduce to the Schrödinger equation. In other words, when physicists speak of a relativistic formulation of some phenomenon, that formulation is supposed to work in the nonrelativistic regime as well. In contrast, a nonrelativistic formulation usually simply falls on its face when extended to the relativistic domain.

Unsuspecting physics students taking a course called advanced quantum mechanics[17] in many universities typically are fed what I call a half-assed treatment of the emission and absorption of photons by the electron in an atom. The electron is treated as a nonrelativistic quantum point particle obeying the Schrödinger equation, while the photon, a relativistic thingy from birth, is not treated as a quantum particle at all. Instead, the quantum electron couples to a classical electromagnetic field obeying Maxwell's equation. The early pioneers of quantum mechanics must have been bothered, as I was as a student, by the unequal status of the electron and the photon in this treatment.

Typical treatment in quantum mechanics textbooks:

electron	quantum	particle	nonrelativistic
photon	classical	field	relativistic

Everybody comes from a field

Consistency and a unified outlook are no less important in theoretical physics than in mathematics; I trust you the reader to go beyond the common and often erroneous view of physicists cranking away just to get some numbers.[18]

The introduction of the Dirac equation provided the first step toward a more equitable treatment: both the electron and the photon could be relativistic. But soon it became apparent that quantizing the Dirac equation while treating the electron as a point particle led to various conceptual difficulties, but perhaps not surprisingly since the first two informative columns of our table were still dangerously unbalanced.

Look at the first informative column. If the electron listens to quantum music, then so must the electromagnetic field. That the world is quantum holds true for all elementary particles.

Indeed, when the electromagnetic field was quantized, as was first tentatively tried by Planck, it was revealed to be a stampede of discrete quanta later named photons, a stampede masquerading as a classical wave under some conditions. But then, how about the electron?

The second informative column remained to be set straight. Eventually, Dirac and others introduced[19] a field for the electron in a conceptual parallel with the electromagnetic field for the photon.

So, after a mighty struggle led by some of the most brilliant minds in 20th century, physicists obtained a "more inclusive and equitable" table:

electron, photon, $\cdots$	quantum	field	relativistic

We took the liberty of including some dots to denote fields yet unknown in the 1930s, such as the wondrous quarks. Compare this table with what is fed to unsuspecting undergraduates. No more two classes of citizens! Everybody is a relativistic quantum field.

Antimatter!

Just as the electromagnetic field can annihilate a photon, the electron field can annihilate an electron. But there is a crucial difference. In contrast to the electromagnetic field, which can create as well as annihilate a photon, we are required, essentially by the conservation of charge, to introduce a "conjugate" electron field to create an electron. When the electron field annihilates an electron, it decreases the charge of the universe by[20] -1, that is, increases the charge of the universe by $+1$. To the contrary, creating an electron would increase the charge by -1.

Since the mathematicians told us that $+1 \neq -1$, the electron field cannot do both. In contrast, the photon is not charged and so the electromagnetic field can both annihilate and create a photon. We are forced to introduce a "conjugate" electron field to create an electron while the electron field creates a particle with charge $+1$.

Voilà the positron, the antielectron! We have arrived at the existence of antimatter.[21] (It would be instructive for the reader to compare this argument with another for antimatter to be given in chapter 7.)

From only particles, to particles and fields, and finally, to only fields

The dear particle known as the electron, who performs so many useful tasks for us in our alleged civilization, is henceforth to be thought of as an excitation in the electron field. Theoretical physicists are mightily pleased: the electron and the photon now have equal rights and are treated equally. As the world turns, eventually each species of fundamental particle is treated as an excitation in a field: a quark field for each species of quark, a neutrino field for each species of neutrino (see chapter 7), and so on. A universe full of quantum fields, surely beyond anything Faraday could have dreamed of!

Think of the present understanding offered by physics of a drop of water containing $\sim 10^{23}$ H_2O molecules. Each electron in that mind boggling multitude of electrons is an excitation in the electron field, each quark an excitation in the corresponding quark field. These excitations are furiously exchanging quanta mediating the strong and electromagnetic interactions: gluons to bind the quarks into the protons and neutrons forming the nuclei, and photons to bind the electrons into atoms. All that frenzy just so that two of these hydrogen atoms could attach themselves to an oxygen atom to form a water molecule. Then a mob of these molecules could form a fluid to nourish life. This fantastic vision of quantum reality is neither a hallucination, nor a feverish dream, but an empirically verified representation of what we call water.

Our understanding of physics has gone through three[22] stages: (1) only particles, (2) particles and fields, (3) only fields.[23]

The issue of identity in the quantum world

Not only are quantum fields entirely real and undergirding our world, they provide an intellectual framework for understanding burning questions about the universe that would otherwise be unanswerable and deeply bother theorists like me. Why are all the electrons identical? And identical means truly identical; indeed, the issue of identity in the quantum world is central to quantum physics.[24] Couldn't the electrons have been made in a factory somewhere, with unavoidable manufacturing defects? Although the tolerance standards are presumably set high beyond any human capability, could we not find some tiny differences between two given electrons if we look hard enough?

Quantum field theory provides an elegantly simple answer to this purely intellectual question. There exists only one electron field, and the zillions of electrons we encounter in the universe are but excitations in this one single field.

The field creates, and it annihilates

In quantum field theory, just as the electromagnetic field can create a photon, the electron field can create an electron, the neutrino field can create a neutrino, a quark field can create a quark, and so on and so forth.‡ This

---

‡As explained earlier, it is actually the conjugate field, but I choose to omit the modifier for the sake of fluency.

view of the world may seem strange to you, and indeed it has struck generations of theoretical physicists as a weird way to describe the goings-on in the subnuclear world of quarks and neutrinos, but that is apparently how the world is constructed.

Enrico Fermi was, as far as I know, the first to take seriously[25] the almost magical ability of a field to create particles by proposing a theory of beta decay in 1933.

I have to digress slightly to tell you about beta decay (the peculiar name is no more than a historical happenstance). Since this book is not about particle "phenomenology" as such, as I said in the prologue, I will be exceedingly brief. First, a nucleus consisting of Z protons and $A - Z$ neutrons,[26] for a total of A nucleons, is traditionally denoted by (Z, A). Second, probably everybody has seen photos of Madame Curie and her fellow pioneers in the lab messing fearlessly with carcinogenic radioactive substances. Through their painstaking experimental work, we now understand that some nuclei are unstable and would "decay" to the "daughter nucleus" $(Z + 1, A)$ by ejecting an electron. (Since the electron carries away 1 unit of negative[27] charge, charge conservation implies that the nucleus is left with one additional unit of positive charge. Thus, Z is compelled to increase by 1, while A remains the same.)

According to Einstein's $E = mc^2$, the energy released in the decay should equal to the product of c^2 and the loss of mass, which is evidently equal to the mass of the "mother nucleus" (Z, A) and minus the mass of the "daughter nucleus" $(Z + 1, A)$ and minus the mass of the electron. Basic accounting, duh. The energy carried away by the electron was dutifully measured, and physicists went crazy: in every case, there was a bit of energy missing, and the missing energy varies from decay to decay.

After a long confusion over the apparent failure of energy conservation (see chapter 5) in beta decay, Wolfgang Pauli hypothesized that the missing energy was carried away by an invisible particle called the neutrino, that is, "little neutron" in Italian, in order to distinguish it from the neutron.[28]

Thus, beta decay is actually the process $(Z, A) \rightarrow (Z + 1, A) + e^- + \bar{\nu}$, with the missing energy carried off by an unseen particle called[29] $\bar{\nu}$, with the Greek letter ν (pronounced nu). (Nowadays, we know that protons and neutrons are made of up and down quarks, denoted by u and d respectively. So beta decay is thought as $d \rightarrow u + e^- + \bar{\nu}$. The down quark d decays into an up quark by emitting an electron and an antineutrino.)

In 1933, the young Fermi boldly proposed his theory: four fields interact at a point in spacetime, one of them being the electron field.[30] Thus, according to Fermi, the outgoing electron in beta decay was not in the nucleus to begin with, but was literally created by the electron field.

Creating the electron: a scandal bringing complete disorder into physics

Nowadays, students of quantum field theory routinely accept that electrons could be created by an electron field. But in 1933, Fermi's theory launched a colossal conceptual breakthrough, which many at the time had difficulty accepting. Senior physicists thought the young turks, Fermi and Heisenberg among them, had gone insane, by claiming that the outgoing electron in beta decay was actually created during the decay process.

Heisenberg recalled: "(I was) criticized very strongly for this assumption by extremely good physicists. I got one letter saying that it was really a scandal to assume that there were no electrons in the nucleus because one could see them coming out; I would bring a complete disorder into physics by such unreasonable assumptions.... It is really difficult to go away from something which seems so natural and so obvious that everybody had always accepted it. I think the greatest effort in the developments of theoretical physics is always necessary at those points where one has to abandon old concepts."[31]

You could literally "see them coming out." So they must have been inside the mother nucleus (Z, A) all along. The younger generation is creating havoc in the house of physics!

Nowadays, nobody blinks an eye. Students are taught that fields create and annihilate, and they dutifully write down the electron field in the so-called interaction Lagrangian (which will be introduced in chapter 10). In reality, to say that the electron shooting out of the nucleus was created out of thin air required a fantastic leap of faith. We revere the greats of physics, such as Fermi and Heisenberg, for leaps such as these.

I feel that, in contrast to some other academic fields, physics is so vibrant precisely because of its vital balance between the young turks and the old fogies, that Fermi and Heisenberg could regard the likes of Planck and Einstein as has-beens. Note however, that Heisenberg referred to them as "extremely good physicists."

Quantum field theory has been tested to far higher accuracy than Newtonian mechanics

Fermi's theory of the weak interaction sounded the first proclamation that quantum fields are real and can create and annihilate particles. In contrast, in quantum mechanics, if you started with seventeen electrons, then no matter how you shake and bake Schrödinger's equation till you are blue in the face, you would still have no more and no less than seventeen electrons.

I began the preface of my recent book *Quantum Field Theory, as Simply as Possible* lamenting a gap in the popular literature on physics:

> By now, there are numerous books introducing quantum mechanics to the general public, but I am not aware of any popular book on quantum field theory. When I told a distinguished theoretical physicist that I was working on a popular book on quantum field theory, he exclaimed, "Your book is really going to fill a gap. By now, everybody and his grandmother has heard about quantum mechanics, but nobody knows anything about quantum field theory." I replied, "Exactly, but even more strangely, by now everybody and his grandmother has heard about string theory." Readers of popular books on physics have jumped directly from quantum mechanics to string theory, it would appear.

Indeed, my motivation for writing *QFT ASAP* was to fill this puzzling gap. Quantum field theory, an ongoing area of research that spans quantum mechanics and a significant portion of string theory, is by far the most advanced and experimentally verified[32] subject in theoretical physics: it has, perhaps surprisingly, been tested to far higher accuracy[33] than Newtonian mechanics and Maxwellian electrodynamics.

Incidentally, the term "quantum mechanics," coined to replace classical mechanics, turns out to be among the least apt terms in physics, since every subject[34] in classical physics can be quantized. Thus classical electrodynamics becomes quantum electrodynamics, known somewhat affectionately as QED. Perhaps the term quantum physics is more apropos, but it is a bit too broad as much of string theory may be regarded as quantum physics. Regardless, let us press on with our story.

Both relativistic, both quantum, and both fields

I told you the story of beta decay to underline for you Fermi's vision of quantum fields creating and annihilating particles. But let us now put beta decay aside and return to atomic physics.

Please admire our attractively egalitarian table displayed a few pages back—a glory of theoretical physics in the 1930s—that presents both the electron and the photon as relativistic quantum fields. Indeed, QED is the simplest realistic[35] quantum field theory that we could study, describing a universe inhabited by only electron and photon fields.[36]

Physicists tackled QED starting in the 1930s, but their calculations often produce results that were infinitely large, leading to much teeth gnashing and hair pulling. Some luminaries even advocated abandoning field theories altogether. So much for luminaries!

As the wheel of history turns, a handful of warmongers in the late 1930s put a stop to all the moaning and groaning among physicists who were suddenly called to more urgent and momentous business. Quantum field theory had to wait till after the war for a bright new generation of brilliant minds, including that of Julian Schwinger and Richard Feynman and many others, to lead it forward into a glorious rebirth.

The young, unlike the aged luminaries, were revolutionary conservatives

Schwinger and Feynman, both twenty-seven in 1945, both lamented that the war had robbed them of the most creative period of their lives.[37] They turned out to be revolutionary conservatives! Instead of heeding their elders' call for trashing quantum field theory, Schwinger and Feynman invented better methods to deal with quantum field theory, notably the fabulous Feynman diagrams.[38,39]

The two stars of postwar American physics were soon engaged in a race to calculate the so-called anomalous magnetic moment of the electron,

The magnetic moment, which we already encountered in chapter 3, measures how a charged particle behaves in a magnetic field, but you don't even have to know that for our purposes here. The electron's magnetic moment has a somewhat tortuous history. By dimensional analysis,[40] it was known to be given by $e/2m$ times a dimensionless number, traditionally denoted by g and expected to equal 1 according to classical physics. Here e and m denote the charge and the mass of the electron respectively.

In 1928, the young Dirac rocketed to fame showing that his equation for the relativistic electron gives $g = 2$, not $g = 1$. The magnetic moment comes out twice[41] as large as what classical physicists would have predicted.

Figure 1. Julian Schwinger's tombstone in Cambridge, Massachusetts. Taken from page 167 of *QFT ASAP*. Wikimedia Commons © Jacob Bourjaily. CC BY-SA 3.0.

But soon, more precise and accurate measurements showed that the magnetic moment was actually a teeny bit larger than the value obtained by Dirac. (The unfortunate historical adjective "anomalous" was tagged on by people expecting Dirac's calculation to be the end of the story. Nowadays we don't think there is anything anomalous about this quantity.)

Historically, the calculation of this teeny deviation from what Dirac had calculated presented the first serious test of quantum field theory. The race was on! It was won by Schwinger, who had the result carved on his tombstone (figure 1).

The strength of electromagnetism is characterized a dimensionless quantity called (for historical reasons) the fine structure constant and defined by[42] $\alpha \equiv e^2/4\pi$. It is measured to be $\alpha \simeq 1/137.036 \simeq 0.007297$, as was already mentioned in chapter 1. (The famed 007 of quantum field theory!) Schwinger's tombstone shows the value[43] he obtained, $\alpha/2\pi$.

Here (figure 2) are the relevant Feynman diagrams, which for our purposes could be thought of simply as pictures showing what is going on.[44] The five diagrams in figure 2, which I will describe here, provide a sketch of what Feynman had in mind when he proposed what he called "funny looking diagrams"[45] that would drive the editors of physics journals crazy.

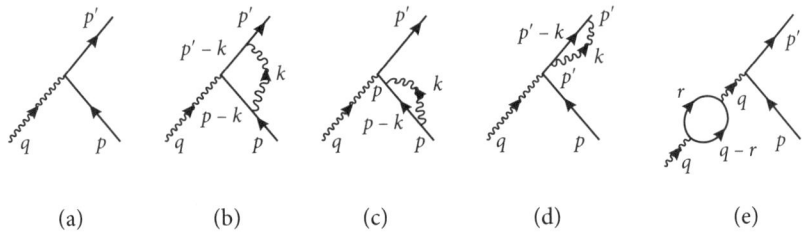

Figure 2. The five Feynman diagrams for calculating the magnetic moment of the electron. Feynman diagrams were invented by their namesake to help himself, and generations of physicists after him, visualize what is going on in spacetime. Adapted from page 196 of *QFT Nut*, Princeton University Press.

These diagrams measure how the electron would behave in a magnetic field[§] an experimentalist might have set up.

In figure 2 (a), an electron carrying momentum p absorbs the incoming photon with momentum q and then continues on its way carrying momentum $p' = p + q$. Dirac's triumph in 1928 corresponds to mastery over this diagram. The other four diagrams in figure 2 were thought to contribute to the anomalous magnetic moment,

Schwinger was the first to be able to calculate diagram (b) in 1948. The electron carrying momentum p first emits a photon with momentum k before it absorbs the photon with momentum q, changing its momentum to $p - k + q = p' - k$. Then it absorbs the photon with momentum k before going on its merry way carrying momentum p'. As you can see, unlike the so-called external momenta (q, p, and p'), the momentum k circulates inside the loop and can in principle take on all possible values: the hard part is to sum up all these possible values of k. (In physics jargon, we have to integrate over k.) Note that in the relativistic quantum world, energy and momentum[¶] add and subtract the same way we learned in elementary schools, and so are still strictly conserved,[46] contrary to what some aged luminaries proposed. Each time there is an interaction, the sum of the incoming momenta has to equal the sum of the outgoing momenta: basic accounting again.

Nowadays, students in a quantum field theory course are expected to be able to calculate these five (actually, only four) diagrams. I will return to diagrams (c), (d), and (e) soon.

[§]The magnetic field's job is to supply the incoming photon carrying momentum q.

[¶]In the relativistic world energy and momentum are intimately related. The word "momentum" is often used generically to cover both concepts.

Renormalization and controlled ignorance

In contrast to their elders, this postwar generation was able to obtain finite and sensible results using quantum field theory. One important byproduct is an understanding of what the infinities that plagued the older generation before the war were all about, a mastery accomplished after much arduous and, dare I say it, heroic work by many theoretical physicists, notably Freeman Dyson, trekking through swamps of misunderstanding and confusion. Much later, this new understanding was synthesized and codified into what is called a Wilsonian view of quantum field theory, in honor of the late Nobel laureate Ken Wilson.[47]

In the Wilsonian view, if we know only the physics below a certain energy scale, call it E^*, we should admit that we are ignorant of the physics above E^*.

You might think that I said this as one of my many jokes, but in fact, this self-evident truth conveys the essence of what theoretical physicists call the renormalization program, started in 1954 by Murray Gell-Mann and Francis Low in a dense and difficult paper.

Instead of describing the framework in general, and so necessarily descending into vapidity, I think it would be better to focus on one specific diagram. Look at (c) in figure 2 and compare with (a). Do you see that (c) has essentially the same structure as (a), except that the incoming electron felt that, before interacting with the incoming photon carrying momentum q, it would be fun to emit a photon with momentum k and then shortly thereafter absorb it?

Let's go back to high school physics for a moment. The notion of mass is basic to physics: Newton thought of it as a measure of inertia, namely, a measure of how much an object resists being accelerated. That notion in its essence is carried over into quantum field theory. The mass of a particle measures the ease with which the particle cruises** through spacetime. By emitting and absorbing the photon, the electron is modifying the ease with which it moves through spacetime. (In the jargon, we say that radiative correction has modified the propagator of the electron field.) In particular, the mass of electron is shifted due to this quantum fun and games of emitting and absorbing a photon.

**The technical term is "propagate." Next time you meet a physicist, ask him or her about his or her favorite Feynman propagator.

How much is the electron's mass shifted? Since in the quantum world, all that is not forbidden by the laws of physics is allowed, the energy and momentum of the photon, which we have denoted by k, could range over all possible values, all the way up to infinitely large values. When pre-war physicists integrated over k, that is, summed up all possibilities, they encountered the bugbear that physicists feared the most, a nasty infinity.[48]

How did the postwar physicists deal with these infinities? Very simple—they denied that there were infinities.

Again, this may sound like a joke, but it isn't. The postwar physicists admitted that they didn't know anything about photons with huge momentum, and so they declined to integrate up to infinite values of k, but stopped when k reached an energy scale called a cutoff and traditionally denoted by the Greek letter capital lambda Λ. Now the shift of the electron mass is perfectly finite and makes total sense, but depends on the value of Λ. In the jargon, it is said to be cutoff dependent. The goal of physics is to determine and measure the cutoff independent quantities (of course).

I call this approach "controlled ignorance": we don't claim to know the physics above the cutoff Λ, but we do want to understand the relationship between the quantities we could measure that does not depend on our ignorance.

Now comes another important point. In the actual QED calculation described earlier in words, the electron's mass is denoted by m. But what is m? It is simply a letter in the Latin alphabet that theorists wrote down on sheets of paper as they calculate. But who cares about that?! What we care about is the electron's mass actually measured by experimentalists. Just to be clear which mass we are talking about, let us call the measured mass m_P with the subscript P standing for "physical." So, schematically, we have a relation like

$$m_P = m + X$$

where X stands for a cutoff dependent quantity due to fluctuations in the quantum fields.

This is known as the mass renormalization of the electron. (Some wits also refer to m_P as the dressed mass and m as the bare mass. The point is that experimentalists never see the bare electron: there is no way to turn off quantum fluctuations.)

March on, you budding quantum field theorist! Look at figure 2 again. You surely could see that (d) is the same as (c) except now it is the turn of the outgoing electron (carrying momentum p') to enjoy some quantum fun and games.

Finally, it is the photon's turn to indulge in some fun, as depicted in (e). As the photon (carrying momentum q) propagates through spacetime, it decides to metamorphosize into an electron and a positron (that is, an electron and an antielectron, carrying momentum r and $q - r$ respectively). Almost immediately after, the electron and the positron annihilate each other, turning back into a photon before it interacts with the incoming electron. This process is known as vacuum polarization[49] (kind of like polarizing a middle of the road population into the left and the right), and as you can see by comparing with (a) it modifies how the incoming photon interacts with the incoming electron. Again, we integrate over r with a cutoff and obtain a cutoff dependent result. Just as with the electron mass, quantum fluctuations force us to replace the letter e, which theoretical physicists have used to indicate the strength of the interaction between charges and the electromagnetic field since time immemorial, by e_P, the charge of the electron that experimentalists actually measure. Again, schematically,

$$e_P = e + Y$$

where Y denotes a cutoff dependent quantity due to fluctuations in the quantum fields (not necessarily equal to X). This is known as charge renormalization.

In summary, in calculating a physical process in QED, theoretical physicists often end up with an answer expressed in terms of m, e, and a bunch of cutoff dependent quantities (which had freaked out the prewar generation who thought they were infinite).

But now, if you eliminate m and e in favor of m_P and e_P, invoking the two relations I just gave you (albeit only in schematic forms) and using just a few lines of high school algebra, lo and behold, in an apparent miracle all the cutoff dependent quantities cancel each other and disappear into (I imagine) the realm of horror movies. We are left with an answer that would make sense to our experimental colleagues and that they could measure, suitable for carving on gravestones.

You can now draw Feynman diagrams with two loops, three loops, and so on ad infinitum. For example, in (b), suppose the photon (with momentum k) emitted by the electron carrying momentum p decided to have the same kind of fun as the photon in (e). Before being absorbed by the electron carrying momentum p', it turns itself into a pair of electron and positron that shortly afterward reconstitute themselves into a photon again. This, as you can see (please draw this!), produces a two-loop diagram. It would hardly surprise you that in the seventy some years since Schwinger's epoch (and gravestone) marking achievement,[50] an army of theorists have calculated diagrams with more and more loops, and a larger army of experimentalists have measured the electron's magnetic moment to higher and higher accuracy, leading to the best agreement[51] between theory and experiment in physics. As I said, quantum field theory has been tested to higher accuracy than classical mechanics.

Two features make this stunning achievement possible.

One is that the electromagnetic interaction, while stronger than the weak interaction, is much weaker than the strong interaction as measured by the small number $\alpha \simeq 0.007$. (Hence the names, strong and weak, of two of the four fundamental interactions.) Feynman diagrams with more loops involve more interaction of the photon with the electron, and thus are proportional to higher powers of α and so are numerically smaller and smaller. In fact, N-loop diagrams are proportional to α^N.

When we calculate diagrams with more and more loops, we expect to encounter more and more cutoff dependent quantities that we will have to beg our experimental colleagues to measure, so we could include them in QED as parameters. But no, by an apparent miracle all the cutoff dependent quantities disappear if we eliminate the bare m and e and express everything in terms of the measured m_P and e_P. I realize that I am being extremely sketchy[52] here. In particular, I do not explain why the miracle just mentioned is only apparent and not a miracle but a rather special property of QED, shared by a large class of quantum field theories, known as renormalizable theories.[53]

In contrast, for nonrenormalizable quantum field theories, as we calculate to larger and larger numbers of loops, what we just expressed as our darkest fear in fact comes to pass, namely that the number of parameters our experimental colleagues need to measure keeps on increasing. As the reader might have heard, quantum gravity[54] is the most infamous nonrenormalizable quantum field theory physicists have been grappling with, for close to a century.[55]

Cutoffs are real: they express our ignorance

I am genuinely puzzled that I still see authors of popular physics books, and even of textbooks, snickering about quantum field theorists surreptitiously sweeping infinities under the rug, but that is based on an antiquated understanding of quantum field theory. No such thing! By the time I got to graduate school, I never heard Schwinger talking about infinities (unless he was referring to an older generation). My mentor Sidney Coleman was careful to say perfectly sensible cutoff dependent quantities when we had to manipulate them, never subtracting one infinity from another infinity.

You muttered, sounding unconvinced, "Physicists are merely replacing the word infinity by the phrase cutoff dependent quantity."

The Casimir force

Let me give an exceedingly concrete example[56] that the cutoff is really there. We saw earlier in this chapter that if the electron is quantized, then the electromagnetic field also clamors to be quantized. Even in a vacuum, with absolutely nothing around, the electromagnetic field is incessantly fluctuating, oscillating with frequencies inversely proportional to the wavelengths of the oscillations (just like waves in the ocean), with wavelengths ranging over all possibilities. In 1948, the Dutch physicist Hendrik Casimir realized that by introducing two perfectly conducting plates into the vacuum we could disturb this incessant fluctuation. See figure 3.

The issue is whether we would be able to observe the consequences of our disturbing the vacuum. As explained in the figure caption, we want to calculate and measure the force between the two plates on the left separated by the distance d.

Now, either from high school science class or from everyday living, we know that the tangential component (that is, the component parallel, but not perpendicular, to the plate) of the electric field must vanish on the surface of the plate. The reason is simple: if the tangential electric field does not vanish, then it would push the electrons in the conductor in the direction of the electric field. But then the excess electrons piled up by the electric field would oppose and cancel the electric field completely, reaching a truce, or an equilibrium as physicists would say. (Textbook statement: In a perfect conductor, an imposed electric field generates a current that tries to oppose the imposed electric field and ends up canceling it. Does that ring a bell? Duh, that's what we mean by perfectly conducting plates: they do not allow tangential electric fields.)

So, the electric field parallel to the plates must vanish at the plates. Thus, the electromagnetic field fluctuating in the vacuum between the two plates realizes that only certain wavelengths are allowed, namely, just those

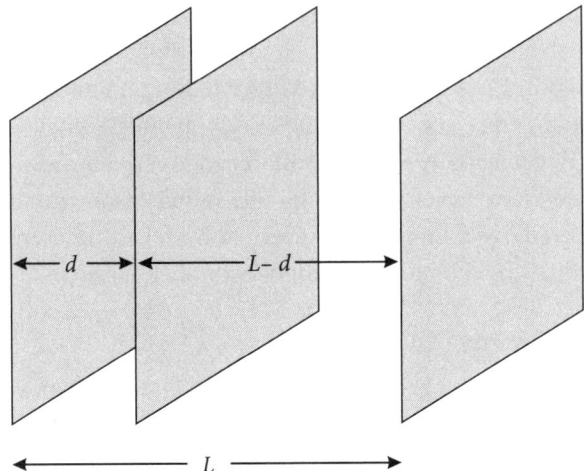

Figure 3. The Casimir effect showing that the vacuum fluctuations of quantum fields are real and measurable. Notice that while I talk about two perfectly conducting plates here, in my textbook I use three plates for technical reasons I don't want to go into, but in the calculation we are going to take $L >>> d$ so that in principle the third plate could be light years away. We plan to keep the two outside plates fixed and move only the inside plate. In other words, by following Casimir, we want to see if there is a force between the two plates on the left separated by the distance d. Reproduced from *QFT Nut*.

whose oscillating electric field parallel to the plates vanishes at the two plates. (Indeed, this amounts in essence to the physics of music. By holding down the two ends of the strings of a violin, the violinist is able to produce sounds of a certain frequency.)

This boundary condition[57] thus compels the half wavelength of electric field fluctuations to be equal to d divided by n, with n an integer $= 1, 2, 3, \cdots$. (Draw the $n = 1$ case to convince yourself.)

The quantum energy in each fluctuation or mode is proportional to the frequency, thus inversely proportional to wavelength, and hence proportional to n. We now add it all up to find the total energy (up to some overall multiplicative constant that does not interest us here)

$$1 + 2 + 3 + \cdots + 999 + \cdots + 999,999 + \cdots$$

Let me now tell what happens next in the form of a classic joke: a mathematician, a physicist, and an engineer walked into the room and looked at the sum we are trying to evaluate. But in a delicious reversal of the usual form of these jokes, the mathematician will be cast as a dumbo and the engineer as the smart guy.

Upon seeing our trying to sum all the positive integers, the mathematician either rushes out of the room screaming "Infinity!" or collapses onto the floor in a dead faint. The engineer says to the physicist, "Perfectly conducting plates[58] do not exist in the real world. If the electric field oscillates too fast, that is, at high frequencies, the electrons simply cannot keep up. Any electromagnetic wave with frequency higher than a certain cutoff frequency would go through the plates, leak out so to speak. So that cuts off the infinite sum. You simply stop after a certain number of terms.[††] Here, I have a handbook that lists what the cutoff frequency is for a copper plate, an aluminum plate, a silver plate, whatever. Just tell me what your plate is made of." The physicist and we agree enthusiastically: "Let's drink to the real world."

I won't show you the rest of the calculation.[59] Suffice to say that if we follow common sense and deal with the sum, not as a mathematical fiction, but as a real life force between two plates that experimentalists can and do measure, we obtain a finite sensible result, as Casimir did more than seventy years before us.

The story I made up about the mathematician, the physicist, and the engineer actually illustrates that what interests each of them may be totally different. The mathematician wants to prove to the other two that summing over all positive integers makes no sense (but as the engineer pointed out to the physicist, that's actually not what we are doing here). The physicist is interested in the cutoff independent[‡‡] and universal energy content of empty space, and doesn't really care about the metal plates except as handy devices to disturb the electromagnetic field. Meanwhile the engineer may be less interested in the universal property of the vacuum than in the cutoff dependent contribution, which may tell her something about how each metal plate conducts electricity at high frequencies. To both the physicist and the engineer, the cutoff is really there.

In fact, the force was first measured by Steven Lamoreaux in 1996, in agreement with Casimir's calculation. As a measure of how feeble the force is, technology took forty-eight years to catch up to theory in this case. I know Lamoreaux, an outstanding and trustworthy experimentalist. I am

---

[††]I am of course simplifying here. In practice, we weigh the subsequent numbers in the sum less and less.

[‡‡]In other words, the theoretical physicists have to show that the force between the two plates that the experimental physicists are to measure does not depend on the cutoff.

sure that he did not sweep anything under the rug and his lab certainly did not blow up in a cloud of infinities.

Parameters in physics

At any given stage, physics is described in terms of a number of parameters, which we are unable to calculate from first principles. They have to be measured by experimentalists. For example, for Schwinger in 1948 e and m are the two parameters of QED. Brilliant though he was, more than enough to calculate for you the magnetic moment of an electron, there was, and still is if he were alive, absolutely no way for him to calculate for you the mass of an electron.

A somewhat peripheral comment that might confuse some readers. As was remarked earlier, in pure QED, with only the electron and photon fields, the electron's physical mass serves to set the scale for mass: any physical quantity with the dimension of mass or energy would have to be expressed as the electron's mass times a pure number. In this simple universe, it is not even meaningful to speak of calculating the electron's mass. Calculate it in terms of what?

But in the late 1960s and early 1970s, the electromagnetic and the weak interactions were unified into a single electroweak interaction, as we will see in chapter 7. The exciting possibility of calculating the electron's mass in terms of the mass of some other particle in the theory arose and caused quite a stir.[60] Unhappily, this possibility was not realized in the actual electroweak theory, but only in some more speculative version of this theory. Of course, we could always hope that, in a not too distant future, the electron's mass will be calculated in terms of something else, starting from first principles. Incidentally, I may note here that Weinberg's last paper was about this tantalizing possibility, ultimately calculating not only the mass of the electron, but the masses of all the quarks and all the leptons including neutrinos.[61]

Effective field theory at low energies or long distances

Closely intertwined with the renormalization program launched after the war is the notion of effective field theory.[62] Again, perhaps the best way to explain this is to refer to a specific example, rather than give a general but vague discussion. To my knowledge, the first substantive use of the effective field theory occurs in 1960 in an influential paper by Gell-Mann and Lévy.[63] They wanted to study the strong interaction* between nucleons (namely protons and neutrons) and pions (the earliest known type of mesons, but once again, as in the discussion about the electron's magnetic moment, it is not necessary for the readers to know all the details in order to appreciate what was achieved).

Suffice it to say that the strong interaction is vastly stronger than the electromagnetic interaction (as was already mentioned) studied by Schwinger, Feynman, and their cohorts, and thus it was generally thought that the approach used by them to calculate the electron's magnetic moment would not work, and indeed it did not. (Schwinger's result immediately yielded the wrong result for the proton's magnetic moment, for which we have to take into account the strong interaction.) Some luminaries even declared quantum field theory dead and proposed a burial with pomp and dignity.[64] But courageously tacking against the prevailing wind Gell-Mann and M. Lévy wrote down an effective field theory (called the sigma model in the professional jargon) consisting of the nucleon and pion fields and incorporating all the symmetries known about the strong interaction at the time.

In all likelihood, the reader has heard that Gell-Mann invented quarks, and that the nucleons and the pions were composed of quarks. This picture led later to the theory of the strong interaction we have now, based on quarks interacting with gluons, and known as quantum chromodynamics or QCD. Well yes, but Gell-Mann hadn't heard of quarks in 1960: he didn't invent them till 1964, and QCD didn't emerge till the early 1970s.

The point is that Gell-Mann and M. Lévy didn't have to know a thing about quarks to write down the sigma model. Yet the sigma model correctly described some of the known behavior of nucleons and pions, by virtue of its incorporating the known symmetries about the strong interaction. The sigma model is an effective field theory, correct at low energies and long distances, of the ultimately correct field theory of quarks and gluons. An important later development was the understanding that an effective field theory could describe two pions scattering off each other at low energies.

In the case of QED, our ignorance of what happens to the electron and the photon at energy and momentum close to the cutoff scale Λ could

---

*More on this interaction in chapter 7.

be summarized in the two quantities m_P and e_P (physicists call them two parameters). In other words, in a universe containing only the electron and the photon, we could calculate any physical quantity involving them, such as for example the electron's magnetic moment, in terms of these two parameters, which experimentalists have to measure for us. Similarly, Gell-Mann's ignorance of quarks and whatnot in 1960 did not prevent him from writing down an effective field theory in terms of a few parameters, such as the nucleon's mass, which had to be taken from experimental measurements.

Another substantive application of effective field theory is to proton decay. We will see in chapter 7 that after physicists understood the strong, the weak, and the electromagnetic interactions by the mid-1970s, they proposed to unify these three interactions into a single grand unified interaction. While grand unified theories are far from empirically established (let alone string theory), they typically predict that the proton, hitherto regarded as the bedrock of stability of our universe, will eventually decay. For example, a possible decay mode would be $P \rightarrow \pi^0 + e^+$, namely a proton disintegrating into a neutral pion and a positron.

Nobody knows whether grand unified theory is correct or not and nobody has ever seen proton decay. Nevertheless, we could write down an effective field theory of proton decay. The idea is simple: regardless of whether the ultimate theory at high energies and short distances allows or does not allow proton decay, the low energy, long distance decay that we observe must respect the symmetries and rules of the low energy physics we have already mastered. For instance, our knowledge of QCD tells us that three quark fields must be involved, and special relativity tells us that four[65] fields are needed, and thus the simplest theory must have the schematic form $\sim qqql$. Here q denotes, generically, a quark field, and l, generically, either the electron or the neutrino field. Remarkably, in order to satisfy the physics we already know, only four terms of the form $\sim qqql$ are allowed. Thus, even in complete ignorance of whether or not the proton decays, and if it does, of the mechanism and theory responsible, proton decay could be described in terms of four parameters for the experimentalists to measure. This approach to proton decay was published independently by Weinberg and by Wilczek and me.[66]

Interestingly, the effective field theory one could write down depends (of course) on what physicists call the "degrees of freedom" relevant for the physics at hand. In particular, if Gell-Mann, years before he conceived of quarks, were to think about proton decay and attempt to write down its effective theory, he would have written the theory in terms of the proton, the pion, and the positron fields, and that theory would be far off the mark.

This effective field theory approach to proton decay is reminiscent of Fermi's theory of beta decay. In 1933, how could Fermi have known about the mechanism and the true theory of the weak interaction? Nevertheless, the constraints of what was known at the time already severely limited what he could write down. In particular, special relativity dictates that four fields have to be involved.[†] Thus, an intriguing ahistorical thought: if Pauli did not propose the neutrino, Fermi would naturally be compelled to invent the neutrino and to make his theory fit, if he truly had faith in a field theory of the weak interaction.

Indeed, we hail Fermi's theory as the first effective field theory, which we anticipate would break down at some energy higher than what was accessible back in 1933. Remarkably, it did break down as anticipated in the late 1960s and early 1970s, when the weak interaction was unified with the electromagnetic interaction. In my decades of wandering around university campuses, I continue to be astounded by learned authorities in other fields pontificating that the theories dear to them would hold across the centuries, even when conditions would have totally changed. The strength of physics is that its theories could reveal their inadequacies and even predict their own demise.

In recent decades, the effective field theory approach has blossomed across several areas of physics.[67] Interestingly, it could also provide an elegant description of the emission of gravity waves by inspiraling black holes.[68]

Physics is sometimes said to be like an onion, to be peeled layer by layer

That theoretical physics is arranged like layers and that we could control our ignorance is striking and surprising. Indeed, this profound conclusion, deduced from centuries of observations, touches upon the mystery underlined by Einstein, that the world is comprehensible. Physics is possible!

[†]That is, an interaction involving only the proton, neutron, and electron fields would not be allowed.

Suppose that to calculate the magnetic moment of an electron in 1948, Schwinger had to know about how the electromagnetic interaction was to eventually wed the weak interaction; his work would have been impossible. Indeed, the courting of the weak interaction by the electromagnetic interaction was initiated by none other than Schwinger himself in the late 1950s and the two interactions were not unified till the early 1970s. Similarly, in 1960 Gell-Mann and Lévy did not have to know a thing about charmed baryons, which were not discovered till the late 1970s, in order to write down their effective field theory. Indeed, charmed baryons and other such exotica do contribute to the magnetic moment of electron, but by a teeny amount that we could estimate. This is yet another aspect of the controlled ignorance practiced in theoretical physics.

To some extent, physics has always been arranged in layers. To study water waves in the 18th and 19th centuries, physicists surely did not have to know about water molecules, but merely a few phenomenological parameters such as density, compressibility, viscosity, and so on. Fluid dynamics is formulated in terms of the velocity of a fluid element $\vec{v}(x, y, z, t)$ at the location specified by (x, y, z) and at the time t. In modern parlance, this would be called a vector field and fluid dynamics could be regarded as an effective field theory of the underlying molecular dynamics, an example of a classical field theory. The fluid element is defined over a spatial region infinitesimally small compared to the length scale characteristic of the phenomenon being studied, such as the wavelength of a water wave, but yet large enough to contains zillions of water molecules.[69]

Quantum field theory, however, makes a much sharper statement. It has a special property first pointed out by Gell-Mann and Low as was mentioned earlier and now known under the general rubric of[70] "renormalization group."

Indeed, that nature is organized in this way is what makes theoretical physics possible. Different energy scales are characterized by different physics. Our ignorance of what goes on about a given energy scale could be summarized in a handful of parameters that we have to ask our experimental colleagues to measure.

I chalk this up as another example of Nature's kindness to physicists. We could easily imagine an alternative universe in which physics could be understood via an "all or nothing" approach. If in order to know some physics, it is necessary to know the physics at all energy scales, physics would have been impossible.

Instead, physics runs on the principle of controlled ignorance, as I just said. A priori, why would the theory that we have to deal with have this particularly pleasing property? A priori, no reason at all. It is one of the deep mysteries of physics. That quantum field theory has precisely this property speaks heavily in its favor.

A never ceasing and ever intertwining dance

In closing, I like to tell a story about the young Einstein. Later in life, he recalled that when he was five years old and sick in bed, his father showed him a compass, and how amazed he was that something invisible could act upon the compass needle through empty space. Thus was the notion of a field impressed upon a young mind, leading to a lifelong fascination. Here I quote J. R. Oppenheimer, writing[71] about Einstein's physics:

> was his total love of the idea of a field: the following of physical phe-
> nomena in minute and infinitely subdividable detail in space and in
> time. . . . It is this tradition which made him know that there had to be
> a field theory of gravitation, long before the clues to that theory were
> securely in his hand.

I would also like to repeat the sentiment I expressed at the beginning of this chapter. I have written a textbook and a popular book[72] about quantum field theory, and so I am painfully aware that one short chapter could hardly do justice to such a basic and essential subject. Those readers who would like to see more about quantum fields are urged to tackle these more detailed treatments.

This chapter summarized: The universe emerges through a never ceasing and ever intertwining dance of quantum fields.

Notes

[1] I say "first" because I didn't get to take a physics course in high school. An article in *Physics Today* listed five Princeton undergraduates influenced by personal contact with Wheeler: James Hartle, David Sharp, Bruce Partridge, Anthony Zee, and Gary Horowitz. See Terry M. Christensen, "John Wheeler's mentorship: an enduring legacy," *Physics Today* 62 (4), 55–59 (2009); see also A. Zee, "A life shaped by John Wheeler," *Physics Today* 62 (10), 10–12 (2009). It is interesting to note that three (JH, AZ, GH) out of the five ended up being professors at the University of California, Santa Barbara. Remember, correlation is not necessarily causality!

[2] Many in the particle theory community share the feeling that the Higgs field will somehow prove not to be fundamental.

[3] As we will see in chapter 7, these forces are but pale vestiges of the electromagnetic force.

[4] I am not sure when gravity was first explicitly recognized as a force. To the ancients, gravity, ubiquitous and ever present, must have been subsumed into a general consciousness of existence.

[5] Perhaps a lesson here somewhere for the young theoretical physicists reading this book. Newton was content to postulate the inverse square law and then explore its consequences. He left its dynamical origin to others like Descartes, whose theory of vortices sweeping the planets along was soon swept into the dustbin of history. I might call the Descartes approach the "all or nothing approach" which some theoretical physicists still indulge in. At any stage in the development of physics, certain questions are not appropriate; for instance somebody could always demand of Newton, "So why inverse square?" A really easy and cheap shot often employed by some audience members during theoretical physics seminars!

[6] A. Einstein, *Out of My Later Years*, Speech at Eighth American Scientific Congress in Washington, DC, on May 15, 1940.

[7] I can't resist saying that the latest and the brashest on the cutting edge of theoretical physics today often seems neither great nor audacious.

[8] For Newtonian mechanics, a knowledge of ordinary differential equations suffices. Partial differential equations are needed only when fluids are involved.

[9] Read about how the earth's magnetic field preserved a memory of when the Babylonians torched Jerusalem in 586 BCE: https://www.timesofisrael.com/burnt-remains-of-586-bce-destruction-of-jerusalem-help-map-physics-holy-grail/.

[10] Ed Purcell, one of the greatest experimentalists of the 20th century and a professor at Harvard when I was a graduate student, drummed on this point, saying that the entire concept of field would be useless were it to depend on how it was produced. Be sure not to read the later edition of his book, which had been mutilated by another author after Purcell's death.

[11] How could it do that? Readers who know some undergraduate level physics could find an explanation in *FbN*, chapter II.2.

[12] Boltzmann was thirteen years younger than Maxwell. While the two greats admired each other's work, they never met.

[13] Heaviside was entirely self taught since his parents could not afford to send him to university.

[14] It has been claimed that the notion and practice of "kidding" is unique to American culture. I tend to agree. In particular, I read that it was totally lacking in England till the 1940s.

[15] A. Garg, *Classical Electromagnetism in a Nutshell*, Princeton University Press, 2012.

[16] In 2011, R. Ahuja et al. (*Phys. Rev. Lett.* 106, 018301 (2011)) discovered that relativistic effects account for about 1.8 volts out of the 2.1 volts produced by the common lead–acid battery. The lead nucleus is so massive that the motion of the electrons around it is highly relativistic. So, the next time you hear a car start up somewhere, you could mutter to yourself, "Ah, Einstein again!"

[17] That sort of physics course sits between an introductory course to quantum mechanics and a course in quantum field theory.

[18] Of course, in the real world that is what almost all physicists do; deep thought won't get you a meal ticket.

[19] This is not a book on the history of physics, but rather a book whose author, by necessity, has to summarize years of confusion by a single sentence. Dirac was apparently just as confused as everybody else about whether the electron should be described by a field. I am indebted to the late Steve Weinberg for an instructive communication on this point. But the Matthews principle has always had a strong hold on theoretical physicists.

[20] Benjamin Franklin, a founding father of the United States and an intrepid kite-flying experimental physicist, was responsible for this arbitrary but unfortunate choice of sign, which endows the basic carrier of electricity with a negative charge, thus bedeviling physics and electrical engineering students ever since with a truckload of minus signs.

[21] For further details, see, for example, *QFT ASAP*, page 155, and *QFT Nut*, page 110.

[22] These three stages remind me of a Chinese martial arts novel that fascinated me deeply in my youth. (1) In the first stage, our hero fights and kills to gain possession of the sharpest sword ever made in the history of mankind. (2) He spends half a lifetime to learn to control the sword using his mind. (3) He dispenses with the sword entirely, burying it in the deep recess of a cave, and disarms his enemies by using his mind only, guiding them on a path toward enlightenment.

[23] I learned recently that, in the American educational system, the authorities in their grave wisdom have introduced something called transitional kindergarten to occur between daycare and kindergarten. (Why?) Perhaps in our quest to understand the universe, the formulation taught in a course on advanced quantum mechanics is akin to transitional kindergarten.

[24] For a detailed discussion, see part VI of *QFT ASAP*.

[25] Older physicists dismissed Fermi by saying in essence that these youngsters didn't know what they were talking about. Same as now.

[26] Remarkably, the neutron was discovered by James Chadwick in 1932, only a year earlier than Fermi's theory. Up till then, it was generally believed that the nucleus was made up of protons and electrons. For a while afterward, some physicists continued to believe that the neutron was a bound state of the proton and the electron, until its mass was measured to clearly exceed the sum of the proton mass and the electron mass. Were the neutron a bound state of the proton and the electron, then according to Einstein its mass would have to be less than the sum of the proton mass and the electron mass since the binding energy would contribute a negative amount to the neutron mass. In fact, the neutron has enough excess mass to decay into the proton and the electron, plus an antineutrino.

[27] As remarked earlier, we're looking at you, Ben Franklin!

[28] A mini time table: 1930 Pauli's proposal of neutrino, 1932 Chadwick's discovery of the neutron, 1933 Fermi's theory of beta decay. The name neutrino was jokingly invented by Edoardo Amaldi in a conversation with Fermi in 1932.

[29] At the risk of being slightly anachronistic, I have written $\bar{\nu}$, following the standardized naming convention used nowadays: the particle emitted in beta decay is now designated an antineutrino, as indicated by the bar and not a neutrino. This is the sort of detail that need not concern us in this book about the intellectual foundation of physics.

[30] In case you are wondering, the four fields are the neutron field, the proton field, the electron field, and the neutrino field. Simple, eh?

[31] *From a Life of Physics*, World Scientific Publishing, 1989, page 48. One argument in favor of the young turks is that, according to Heisenberg's uncertainty principle, given the size of the nucleus ~ 1 fm, the uncertainty principle tells us that the typical momentum for an electron whose position was hitherto confined to the nucleus should be $\sim \hbar c/1$ fm ~ 200 MeV. This is typically much higher than the momenta of electrons emitted in the decay.

[32] String theory remains to be experimentally verified, as almost everyone knows.

[33] See page 2, *QFT ASAP*. For the goof in the *New York Times*, see page 242.

[34] Even gravity, although the result tends to go out of control at short distances.

[35] That is, not a "toy model" some theoretical physicists play with for their amusement and edification.

[36] In this simple universe, the electron's physical mass serves to set the scale for mass. For example, due to their mutual electromagnetic attraction, a positron and an electron could orbit around each other and form what is called a positronium. The positronium's mass would be proportional to the electron's mass.

[37] They both did military research, on radar and on the atomic bomb respectively.

[38] Which in spite of what the uninformed Feynman idolaters might think, are now, so many decades later, finally showing their inadequacies. See, for example, page 484, *QFT Nut*.

[39] Read, on page 170 of *QFT ASAP*, about how, even well into the 1960s, Schwinger's students at Harvard had to hide in order to draw Feynman diagrams.

[40] It would take us way too far afield to go into that here. For interested readers, see *FbN*, chapter 1.

[41] One of the most famous factors of 2 in the history of physics. I show how it pops up in equation (5) on page 195 of *QFT Nut*.

[42] In the Heaviside-Lorentz units used in quantum field theory with $\hbar = 1$ and $c = 1$. See, for example, page 393 of *FbN*.

[43] Incidentally, this perturbative approach is even more successful than one might think because subsequent terms are further suppressed by powers of π.

[44] For a detailed exposition, see *QFT Nut*.

[45] For the spread of Feynman diagrams in postwar physics, see D. Kaiser, *Drawing Theories Apart: The Dispersion of Feynman Diagrams in Postwar Physics*, University of Chicago Press, 2005.

[46] See chapter 2.

[47] See B. E. Baaquie et al., "Ken Wilson Memorial Volume: Renormalization, Lattice Gauge Theory, the Operator Product Expansion and Quantum Fields," WSPC, 2015. You might be interested in the essay by yours truly about my encounters with Wilson.

[48] The reader might be aware that there are different degrees of infinity. Victor Weisskopf, later the eminence grise of MIT, discovered that the infinite shift in the electron mass is the mildest form of infinity.

[49] I heard that vacuum polarization was what Feynman had trouble with in his losing race with Schwinger. (But remember, I am not a historian.) The enormous irony is that we now know that to determine the anomalous magnetic moment of the electron it was not necessary to calculate (e)! By the way, the calculation is indeed rather tricky. See, for example, *QFT Nut*, chapter III.7, "Polarizing the Vacuum and Renormalizing the Charge." Some readers might realize that the vacuum acts like a dielectric material, as in electrical engineering.

[50] Once when I lectured about quantum field theory to a general audience in Bariloche, a picturesque Argentine town nestled in the foothills of the Andes, I showed a photo of Schwinger (the same one as on page 170 of *QFT ASAP*). After the lecture, a pleasant young Argentinian came up to me and asked me if I noticed anything unusual. I didn't, but then he revealed that the glasses he was wearing had exactly the same frame as Schwinger's. I found this charming and refreshing since in the United States the number of Feynman worshippers must outnumber those who have heard of Schwinger by at least 10 to 1.

[51] Of course, it would be exciting if a discrepancy emerges, which would herald some hiherto unknown physics. See *QFT ASAP*, page 2 and pages 239–242, which by the way documents one of the most embarrassing mistakes of physics reporting in the *New York Times*.

[52] Readers who have the appropriate background and would like to see more details will find them in various modern textbooks on quantum field theory. For instance, I devote part III of *QFT Nut* to explaining in some detail what you just read here. Other readers will have to make do with my semipopular book *QFT ASAP*.

[53] The simplest realistic quantum field theory that we could study, namely QED containing only the electron and the photon field, still entertains a wealth of physical phenomena, such as electron-electron scattering, electron-positron scattering, and photon-electron scattering. Using Feynman diagrams, you could calculate each of these processes to however many loops you like. The claim is that upon eliminating the unphysical e and m in favor of e_P and m_P we could always make the cutoff dependent quantities go away.

[54] For some heuristic thoughts about quantum gravity, see, for example, chapter X.8 of *GNut*.

[55]One widespread misunderstanding is that if we do not have a complete theory of quantum gravity, valid for all energies, then we cannot compute how gravity behaves in the quantum world. But that can't be true: just from the uncertainty principle we know that the distance r between two masses is subject to quantum fluctuations and so Newton's law must be modified. It has been shown that the potential energy between two masses is modified to $V(r) = -(GM_1 M_2/r)(1 + 41G\hbar/10\pi c^3 r^2 + \cdots)$. See N. E. Bjerrum-Bohr, J. F. Donoghue, and B. R. Holstein, arXiv:hep-th/0211072; J. F. Donoghue, arXiv:gr-qc/9405057. The correction to Newton's expression is miniscule, but definitely they are perfectly calculable.

[56]Taken from my book *QFT Nut*, chapter I.9.

[57]For readers who remember their high school trigonometry, the electric field is proportional to $\sin \frac{\pi x/2}{d/2n} = \sin \pi nx/d$, for n a positive integer, and where x denotes the direction perpendicular to the plates. This expression vanishes at $x = 0$ or $x = d$, as it should. Some sophisticated readers might note that I am talking about the simplified version presented in *QFT Nut* in which I omit the two transverse directions so as to avoid doing an integral over k_y and k_z.

[58]This naturally reminds me of a classic joke. A beautiful woman bought a gorgeous convertible and invited a perfect man for a ride in the countryside. They were enjoying the summer sunshine and the winding country road when suddenly they saw, surprise, Santa Claus walking along. The woman swerved hard, narrowly missed killing Santa Claus, and plowed into a tree. The car was destroyed but fortunately nobody was hurt. The woman got out of the car and let loose with a few choice curse words. Question: How many people survive this unfortunate accident?

Answer: One, since the perfect man, just like perfectly conducting plates and Santa Claus, do not exist.

[59]Those readers who want more details could consult my quantum field theory textbook, *QFT Nut*, chapter I.9.

[60]S. Weinberg was among those who raised this possibility. My doctoral student Steve Barr worked on this with me and published a few papers.

[61]Weinberg sent me the manuscript (later published in *Phys. Rev. D* 101, 035020 (2020)) he was working on. I dutifully read through it, and told him that his references were off: the paper had only three references and two of them are to Barr and me (see previous endnote). I told him that was rather irregular since Barr and I were inspired by his paper. Ultimately, he did reference his own paper from the 1970s in the published version. See also my essay "In memoriam: Steven Weinberg (May 3, 1933–July 23, 2021)" in *Nuclear Physics*, section B, vol. 998, 116411 (January 2024). https://www.sciencedirect.com/science/article/pii/S0550321323003383?via%3Dihub.

[62]See chapter VIII.3, "Effective Field Theory Approach to Understanding Nature," in *QFT Nut* for an introduction. For an overview, see S. Weinberg in *Ken Wilson Memorial Volume: Renormalization, Lattice Gauge Theory, the Operator Product Expansion and Quantum Fields*, edited by B. E. Baaquie, K. Huang, M. E. Peskin, and K. K. Phua.

[63]M. Gell-Mann and M. Lévy, *Il Nuovo Cimento*, vol. 16, 1960, pages 705–726. For a textbook overview, see *QFT Nut*, chapterVI.4.

[64]For the near death and glorious resurrection of quantum field theory, see, for example, chapter V.2 of *QFT ASAP*.

[65]Fermion fields.

[66]S. Weinberg, *Phys. Rev. Lett.* 43, 1566, 1979; F. Wilczek and A. Zee, ibid., 1571. For a pedagogical discussion, see *QFT Nut*, pages 455–456.

[67]For its appearance in condensed matter physics, see parts V and VI of *QFT Nut*.

[68]See chapter N.1 of *QFT Nut*.

[69]I teach a course on fluid dynamics, and the physics of water waves offers an interesting parallel to particle physics. As the wavelength decreases, "new" physics emerges. For example,

at a scale of a few millimeters, waves turn into ripples. Surface tension, which is a direct manifestation of molecular cohesion, then takes over. See *FbN*, chapter VIII.3.

[70] A detailed explanation would be far beyond the scope of this book; see, for example, *QFT Nut*.

[71] The *New York Review of Books*, March 17, 1966, issue, published less than a year before Oppenheimer's death.

[72] Perhaps one and a half; I was thinking of *QFT ASAP*, but a good portion of *Fearful* is about field theory.

5
FEARFUL SYMMETRY: A UNIVERSE FULL OF SYMMETRIES

Symmetry and the search for beauty in physics

I would like to tell you about conservation and symmetry, two central concepts in theoretical physics, and the unexpected connection between them.

The universe commands: thou shalt conserve

If you cannot find your key, you know that it must still be somewhere in the universe. It could not have vanished into thin air. Matter is conserved. In our daily life, we invoke conservation laws constantly without consciously thinking about it.

Physics has established that all sorts of quantities in the universe are conserved. Energy is conserved. Momentum is conserved. Mass is conserved. Electric charge is conserved.*

To be a bit more precise, when physicists say that energy is conserved, they mean that if we add up the energy of all the particles going into any physical process, this sum must equal the sum of the energy of all the particles coming out. Not one whit less, not one whit more. There is no free lunch in physics. Energy cannot appear out of nowhere nor disappear into nowhere. What you get out is what you put in, and the accountant in the sky never slips up.

Conservation is the law of the realm.

Everyday physics, just like everyday life, appears to be full of loss and dissipation. But that is only apparent. Throw a marble into a bowl. Come back later, and you would be astonished if it is not resting serenely at the

---

*This is not meant to be a complete list of conserved quantities in physics.

bottom of the bowl. The kinetic and potential energy[1] possessed by the marble as it left your hand has diminished to a small amount of residual potential energy without any accompanying kinetic energy. But in fact the disappearance of energy is illusory because we choose to ignore other forms of energy. By rattling in the bowl, the marble has generated sound and heat, both of which escaped into the environment, thus increasing the entropy of the universe. (See chapter 9 for more about entropy.)

As physics progresses, some of these conservation laws merge. At one time, physicists thought that energy and mass were separately conserved (and they still are in everyday life). But Einstein with his $E = mc^2$ famously taught us that mass could be converted to energy and vice versa. Roughly speaking, we could regard mass as a form of energy so that two separate conservation laws are combined into one. (Effectively there are still two in everyday life, even though deep down there is only one conservation law, because the mechanisms converting energy and mass to each other are typically not available, unless you are operating an accelerator or setting off atomic bombs in your neighborhood.)

I need hardly remind you that physics is an empirical science. These laws did not come to physicists as divine revelations, but were painstakingly established over the centuries through careful experiments, with many setbacks and confusion. For instance, in our marble in a bowl example, the tiny amount of energy carried away by sound and heat has to be carefully trapped and measured. Historically, it has not been smooth sailing for some conservation laws either. For instance, as was alluded to in chapter 4, before Pauli came up with the idea of a hitherto unknown and undetectable particle carrying away energy, various eminent theorists championed the suggestion that the "sacred law" of energy conservation might fail in nuclear decay. We of course now know, in the glare of hindsight, that this isn't so and that energy conservation still stands in nuclear decay. But at the time, with all the weird stuff going on in the quantum domain, this suggestion did not seem so outlandish at all.

Where do the conservation laws come from?

Some conservation laws simply pop out of Newton's equation of motion during an introductory physics course. For example, when a planet goes around its sun in an elliptical orbit, it sometimes moves fast, sometimes slowly, sometimes far from the sun, sometimes near, but yet, as long as the planet obeys Newton, the sum of its kinetic and potential energy does not vary in time.[2]

Since these conservation laws also apply when two quarks smash into each other to produce jets of gluons, they clearly must transcend a specific equation of motion and have a deeper meaning. Indeed, energy conservation holds for all four fundamental interactions we know about (and that we will meet in chapter 7), even though the four differ enormously in their various manifestations. So, there must be something profound behind these conservation laws. What could that be? Not some wizard behind the curtain making sure that nothing ever goes missing in the universe.

For those readers who took high school physics or perhaps some courses beyond that: Did your professor ever say where these conservation laws "come from"? I am here to tell you the amazing news, in case you missed it, that we now know where. Old news, but still amazing, indeed almost beyond belief.

Symmetries in the universe

When you first learned to crawl, your brain knew enough to exert equal amounts of force on your left and on your right side. In fact, with few exceptions, all animal life forms on earth, and plausibly all lifeforms in the universe, who want to locomote straight ahead are constructed with bilateral symmetry. Internal organs, in contrast, do not face the same evolutionary pressures and could be placed left, right, or center. Similarly, while cars are externally left-right symmetric, on which side the driver sits is a matter of national convention.

Thus, the idea of symmetry comes naturally to physicists, as we have already seen in chapter 2. Newton understood that the apple did not fall down, but fell straight toward the center of the earth. There should be no special direction in space.

Symmetry demotes the special.

The appreciation and the discovery of ever larger symmetries has become a central theme of 20th century physics. A detailed discussion of this important subject merits an entire book, and entire books have been written, including my first popular physics book, *Fearful Symmetry*. See figure 1. Here I could merely convey to you that the deeper we explore, the more symmetrical the laws look. (I will come back to this in chapter 7.)

Symmetry is fundamental to physics

Consider the circle and the irregularly shaped polygon shown in figure 2. Imagine a psychiatrist showing a hapless subject these two geometrical figures in one of those silly and possibly meaningless association

KORKUNÇ SİMETRİ

MODERN FİZİKTE GÜZELLİK ARAYIŞI

Türkçesi: *Mustafa Bayrak*

3

KETEBE
BİLİM

ROGER
PENROSE'UN
ÖNSÖZÜYLE

A. ZEE

Figure 1. The Turkish translation of *Fearful Symmetry*.

tests. If the subject is a high school student struggling in his physics class, and the psychiatrist asks him which figure he associates with physics, he would likely pick the irregularly shaped polygon. The sharp corners poking out randomly are associated with pain in his mind. (Aha, deep insight into human psychology!) But suppose that the subject is a physicist asked to associate one of these two figures with the fundamental laws of the universe. She will no doubt pick the circle. If Nature is indeed full of irregularities at the fundamental level, quite a few theoretical physicists, certainly myself included, would quit and go study something simpler.

The basic belief is that the fundamental laws of physics should be simple and elegant. Over the last couple of centuries, as physics advanced, the astonishing discovery is that this expectation holds true. The deeper

Figure 2. Which of these two figures is simpler and more symmetrical?

we look, the more symmetrical the view, as I have already said. When theoretical physicists say something is simpler, they typically mean more symmetrical. Thus, in figure 2, the circle is simpler than the nasty polygon.

During the 20th century, symmetry has emerged as a central concept of theoretical physics. But before I can talk about symmetry in physics I have to tell you about a branch of mathematics called group theory.

Group theory blew the "old fogeys" away in the 1930s

Allow me to introduce you to Eugene Wigner, who won the Nobel prize for applying group theory to quantum physics. He recalled,[3]

> The older generation, including Einstein, Schrödinger, and Pauli, looked at group theory with dismay. Pauli coined the term die Gruppenpest (that pesty group business). Schrödinger told [me], "But surely no one will still be [using group theory] in five years." ... I relayed my doubts to Jancsi von Neumann, who reassured me: "Oh, these are old fogeys. In five years, every student will learn group theory as a matter of course."

The old fogeys like Schrödinger were totally wrong. I now teach group theory to undergraduates regularly. (Preview: We will meet the venerable Wigner again in chapter 8.)

The concept of symmetry appears to be innate in the human mind, and emerges naturally from geometry. We readily understand that the equilateral triangle is more symmetrical than the isosceles triangle. See figure 3.

Symmetry and the notion of transformation

This intuitive sense of symmetry could be quantified by enumerating the transformations that leave a geometrical figure unchanged. Thus, the isosceles triangle is left unchanged, or invariant, when we reflect it about

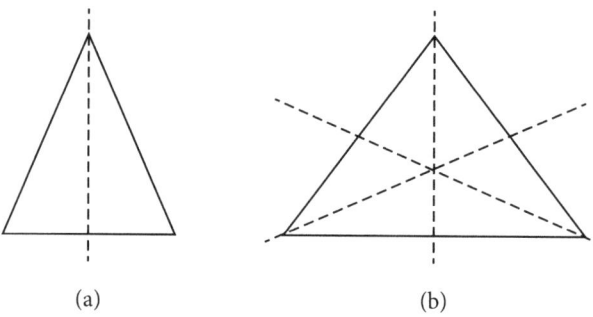

(a)   (b)

Figure 3. Is the equilateral triangle or the isosceles triangle more symmetrical? In the isosceles triangle, one of the three sides is "special"; in the equilateral triangle, none of the three sides can claim to be special. Symmetry is the demotion of the special.

its median to the base, indicated by the dotted line in figure 3(a), or to use everyday language, when we flip it over. The equilateral triangle, on the other hand, is more luxuriously endowed with three identical medians, as indicated by the three dotted lines in figure 3(b), and it is left invariant when reflected about any one of these three medians. Furthermore, it is also left invariant by rotations through 120° and[4] 240°.

Mathematicians simply list the transformations that leave a geometrical figure invariant, and call the set (or group) of all such transformations the invariance group of that figure. More jargon: The transformations are referred to as elements of the group. The more elements the invariance group of a geometrical figure possesses, the more symmetrical it is. In our example, the invariance group of the equilateral triangle boasts of six elements[†] compared to the paltry two[†] elements in the invariance group of the isosceles triangle. This quantifies the intuitive sense that the equilateral triangle is more symmetrical than the isosceles triangle. Similarly, to mathematicians and to physicists, the hexagon is more symmetrical than the equilateral triangle.

I trust that you could now see the direction in which mathematicians are going. The invariance group of an N-sided regular polygon[5] gets to be ever larger as N increases.

[†]Why six and two? Just in case you are going crazy, having counted only five and one respectively, let me assure that you counted right. Crucially, mathematicians also include the so-called identity transformation, known in everyday language as "doing nothing." I say crucially because this is akin to the importance of introducing the number 0, aka nothing, in the law of addition.

You could also see that the invariance group of the circle is qualitatively different from the invariance group of an N-sided regular polygon. The circle is left invariant by rotations through any angle θ. The invariance group of the circle is known as a continuous group, in contrast to the invariance group of the N-sided regular polygon, which is known as a finite or discrete group.[6] The N-sided regular polygon is left invariant by rotations through a discrete set of angles, namely multiples of $360°$ divided by N. In contrast, the angle θ of a rotation that leaves a circle invariant could take on any value, as was just said.

The notion of multiplying transformations[7] together[‡] arises naturally. We could perform two transformations one after another. The resulting transformation is known as the product of the two transformations. For example, if we first rotate the circle through $17°$, and then rotate it through $21°$, that is the same as rotating the circle through $17° + 21° = 38°$. So the invariance group of the circle, known as $SO(2)$, is associated with addition, except that when you get to $360°$ you go back to $0°$, assuming that you are a Babylonian at heart. (If you are a modern person, who prefers radians to degrees, then it is addition modulo 2π, which is merely a fancy way of saying that in radians the angle 2π is the same as the angle 0.)

So there you have it, a lightning introduction to group theory. I am reminded of the joke of a mathematician explaining to her parents what she does. They asked in disbelief, "From this you make a living?"

Enchantment and piety

Well, things become more complicated and considerably more interesting when we move up to 3-dimensional space. Even something as apparently innocuous as rotations displays an unexpected behavior. Consider the group consisting of rotations in 3-dimensional space, known as[8] $SO(3)$. When we multiply together two rotations, call them R_1 and R_2, we find that $R_1 \times R_2$ is not equal to $R_2 \times R_1$, as you can see in figure 4. The order matters, unlike ordinary multiplication for which 3×5 is for sure the same as 5×3. To test if you have absorbed this fundamental fact about the space

---

[‡] Ordinary multiplication could also be thought of in terms of transformation. Imagine "transforming" a photo by enlarging it by a factor of 3. Then enlarge it by a factor of 5. The result is that the photo has been enlarged by a factor of $15 = 5 \times 3$. Again, I am blurring over technicalities. For more details, see *Group Nut*.

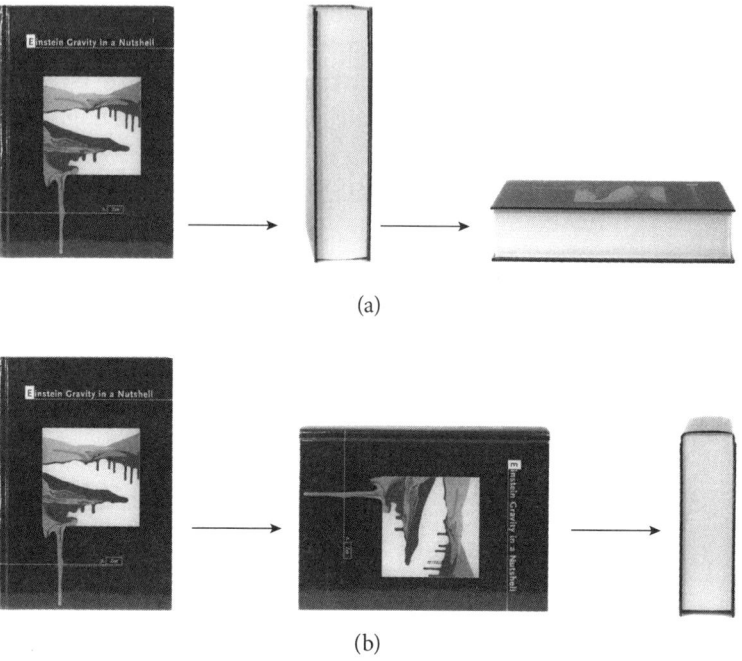

Figure 4. Consider rotating a book, first by 90° around the z-axis, pointing up, and then by 90° around the x-axis, pointing out of the page. The result is shown in figure (a). Perform the same two rotations but in the opposite order. The result is shown in figure (b). Rotations do not commute. Reproduced with permission from *Group Nut*.

we inhabit, I cordially invite you to rotate the object nearest to you, namely your body.[9]

Reversing the order of the two entities you are multiplying is known in the jargon as commuting them. (They each travel to where the other guy lives?) So, think of a group as a mathematical structure in which the order of multiplication matters. The rotational group $SO(3)$ is said, in the jargon, to be noncommutative or nonabelian. Multiplication of group elements in general do not commute.[10] (In contrast, the group of rotations in two dimensions $SO(2)$ is said to be abelian. When you perform 2 rotations in 2-dimensional space in succession, the rotation angles simply add, as was noted earlier.)

Humans could be divided into two sets. In our post-Enlightenment age, the vast majority might grumble, "Meh, who cares?" A small minority, when shown that rotations do not commute, might exclaim, "How fascinating!" and if young enough, might even go on to become mathematicians, physicists, and various citizen intellectuals of all types. Since you are reading this book, you probably belong to this curiosity driven minority.

Of this small minority, some who actually studied group theory have experienced the enchantment and pious awe the subject evoked in Wigner, as he reminisced in the following quote. "Soon I was lost in the enchanting world of [group theory]. . . . The 1905 article by Frobenius and Schur [that von Neumann showed me] was my primary introduction to [group] representation theory, and I was charmed by its beauty and clarity. I saved the article for many years out of a certain piety that these things create."[11]

It is never too late for you to experience the magic of group theory. Go tackle a textbook[12] on group theory. I have experienced what Wigner was talking about.

Beauty and clarity, enchantment and piety, those are not the sort of words most of our fellow humans would associate with mathematics and physics!

Discrete versus continuous symmetries

For the rest of this chapter, I will discuss only continuous symmetries such as rotations. While discrete symmetries play an absolutely essential role in physics, because they do not fit into the main theme of this chapter, I am leaving them out with enormous reluctance. I will merely list the three most important discrete symmetries.[13]

Parity: The laws of physics in the mirror world are the same as in our world.
Charge conjugation: The laws of physics remain the same when we flip matter into antimatter.[14]
Time reversal: When we make a movie of a microscopic physical process and run it backward, the audience cannot tell, using physics alone,[15] whether the movie is playing forward or backward in time.[16]

One bombshell development in the physics of the second half of the 20th century was the discovery, much to the shock and dismay of physicists, that the laws of physics violate all three of these symmetries.

While these three discrete symmetries are of fundamental importance in physics, the invariance groups associated with them are mathematically almost trivial. In fact, they are the same as the invariance group of the isosceles triangle (known as Z_2 if you would like to know) because all three symmetries correspond to flipping one side of the isosceles triangle to the other: for parity, flipping our world to the mirror world; for charge conjugation, flipping matter into antimatter; for time reversal, flipping a forward running movie into a backward movie.

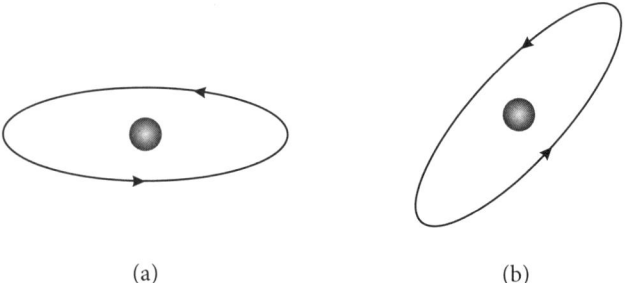

(a)   (b)

Figure 5. (a) The elliptical orbit followed by a planet around the sun. (b) The orbit in (a) rotated by 47°.

Seemingly a branch of mathematics that physicists could not use

A classical physicist, let's say Newton, when shown group theory, might say, "Interesting, but what could it do for physics?" The answer is not that much for classical physics.

Let's see why. Again, consider the Newtonian problem of a single planet following an elliptical orbit around the sun. Given an orbit as shown in figure 5(a), rotate it by 47° as shown in figure 5(b). (We actually have to specify three angles[17] in order to specify a rotation in 3-dimensional space. For the sake of keeping the discussion simple, I will just say one angle instead of three.)

In Newtonian physics, rotational invariance merely tells us that the rotated orbit is also a possible orbit. Big deal, that is totally obvious! No need for group theory.

Indeed, as group theory was being developed in the late 19th century, a mathematician bragged that mathematicians had finally invented something that physicists would not be able to "steal" and use. But he had no way of anticipating the advent of quantum physics in the 20th century!

Quantum physics exalted group theory

Consider the hydrogen atom, consisting of an electron going around a proton, superficially like a planet going around a star. But quantum mechanics is fundamentally different from classical mechanics. We can no longer talk of the electron tracing out an orbit.[18] Instead, we have a wave function $\psi(x)$, something like a cloud that determines the probability amplitude of locating the electron at the point x. As was indicated in chapter 3, the state the electron could be in is quantized. For instance, for specified amounts of energy and angular momentum allowed by quantum mechanics, there are only five possible wave functions (corresponding to states[§]) for the electron, which for the sake of definiteness we list here as ψ_{+2}, ψ_{+1}, ψ_0, ψ_{-1}, ψ_{-2}, without saying what the notation means.[19] The probability that the electron is located at x is given by the absolute square of the wave function (as was also mentioned in chapter 3). Just to be concrete, let me show you what the probability clouds look like. See figure 6.

Carved in stone

The important point of our discussion, however, is not the shape of these electron clouds, which I am showing you just for the sake of definiteness, and for fun, so to speak. The important point is how these wave functions behave under rotations.

Earlier, we discussed what happens when we rotate a Newtonian orbit by 47°. Well, nothing much. We just get the same orbit tilted by 47°. But a huge difference arises when orbits are replaced by probability clouds. Consider the wave function ψ_{+2} rotated by 47°. Call the result ψ'_{+2}. But this is not in the list of the five wave functions allowed by quantum mechanics. So ψ'_{+2} must be expressible as a sum[20] of the five listed wave functions, with five arbitrary coefficients c_{+2}, c_{+1}, c_0, c_{-1}, c_{-2}, namely $\psi'_{+2} = c_{+2}\psi_{+2} + c_{+1}\psi_{+1} + c_0\psi_0 + c_{-1}\psi_{-1} + c_{-2}\psi_{-2}$. This is known as a linear combination[21] of the five listed wave functions. (Incidentally, you have already seen in chapter 3 that quantum states could be linearly superposed or combined.[22])

The triumph of mathematics is that these five coefficients c_{+2}, c_{+1}, c_0, c_{-1}, c_{-2}, each one of which depends in some intricate fashion on the rotation we are performing, are completely determined by group theory!

In other words, the bragging and anti-theft mathematician in the 19th century, without the slightest inkling of quantum mechanics, could have told us what these five intricate functions of rotations were. They are completely fixed by the structure of the rotation group $SO(3)$ (in particular, that rotations do not commute). As alluded to in Wigner's reminiscences, this is known as representation theory in mathematics. The five wave

---

[§]For our purpose here, we use the terms wave function and state interchangeably.

3d

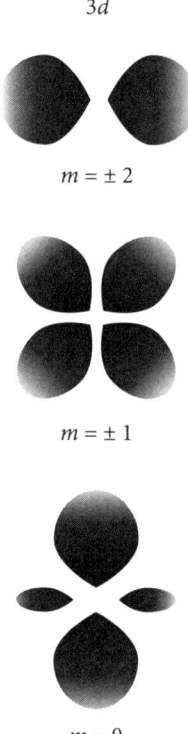

$m = \pm 2$

$m = \pm 1$

$m = 0$

Figure 6. The probability distribution of an electron in a particular quantum state in a hydrogen atom. It turns out that $|\psi_{+2}(x)|^2 = |\psi_{-2}(x)|^2$ and $|\psi_{+1}(x)|^2 = |\psi_{-1}(x)|^2$ and so there are only three, not five, distinct probability clouds, as are shown here. (The subscript on $\psi(x)$ is labeled m in the figure; thus $m = \pm 2, \pm 1$, and 0.) Adapted from Leighton, page 178.

functions ψ_{+2}, ψ_{+1}, ψ_0, ψ_{-1}, ψ_{-2} are said to furnish a 5-dimensional representation of $SO(3)$. In this particular example, the number is five. It cannot be four and it cannot be six. The five is carved in stone, fixed once and for all.[¶] This is important for our discussion later.

The remarkable point is that a hypothetical mathematician who has somehow never heard of rotation and angular momentum would still be able to tell us the dimension of the representations of $SO(3)$.

[¶]This might ring a bell. Recall that in chapter 3 I said that angular momentum is quantized to have the value $l = 0$, 1, 2, 3, $\cdots$. This quantization is indeed related to the dimensions of the representations allowed by the group $SO(3)$, namely the sequence $d = 2l + 1 = 1$, 3, 5, 7, $\cdots$. These numbers are carved in stone, as I emphasized in the text.

Executive summary:

1. How a quantum wave function (or more precisely, a quantum probability amplitude) behaves when transformed under a symmetry group is completely determined by group theory.

2. The number of quantum states with a specified amount of energy and angular momentum allowed by quantum physics is completely determined by group theory.

Symmetry of the laws of physics, not the symmetry of the situations

In elementary physics, students might be asked to solve the electric field around a charged sphere, or around a charged cube. The relevant symmetries in the two homework problems are manifestly different, spherical in one case and cubical in the other. The symmetry is simply due to the situation the electric field finds itself in. It is important to underline that in this book we are not talking about the symmetry, or the lack thereof, imposed by the situation under consideration, a charged sphere in one homework problem, a charged cube in the other.

Instead, our interest here is in the underlying symmetry of the physical law, namely the law governing the electric field in this example. And electromagnetism, as governed by Maxwell's equations, is explicitly rotationally invariant. Maxwell's equations do not pick out a special direction, any more than Newton's equation for gravity picks out a special direction. (Remember, the apple does not fall down, but it falls toward the center of the earth!)

To put it bluntly, electromagnetism couldn't care less about the shape of the charged object the instructor wishes to torment the students with. A sadistic instructor could ask the students to work out the electric field around a charged object in the shape of an octopus and the universe wouldn't care. That electric field would certainly not be rotation invariant, even though the laws of electromagnetism surely are.

The important point here is to distinguish clearly between symmetry of the laws and symmetry of the solutions of the laws pertinent to a given situation. It may sound like a minor distinction, but in fact it was, and is, huge. By confounding the two, the ancient Greeks committed a grave error. They understood that the laws governing the motion of the planets should not favor any particular direction, in other words, should be spherically symmetric, but that does not mean that the planets must orbit in circles. When observations showed that planetary orbits are not perfectly circular, Ptolemy and his school of astronomers had to add epicycles upon epicycles. Missing the distinction between laws and solutions could indeed retard science for close to a couple of millennia.

By the same token, the probability clouds are not all round balls in quantum mechanics.

The deeper we go, the more symmetric the laws look

The history of theoretical physics over the last century or so could be summarized, with considerable, or even outrageous, exaggeration, as the search for and the discovery of ever larger invariance groups in the fundamental laws of physics.[23] Clearly, I do not have the space to devote even the barest amount of explanation to these groups. Indeed, each of these groups is associated with one or more physics Nobel prizes and merits entire books detailing what it is and how its relevance to physics was realized theoretically and verified experimentally. So, please regard the following as merely a list to whet your appetite and to entice you to learn more about them.[24]

From $SO(3)$ to $SO(3, 1)$

First, the granddaddy of these discoveries: Einstein and others showed around the turn of the 20th century that electromagnetism enjoys a larger symmetry group than mere rotation, as we will discuss in chapter 6. Indeed, using modern mathematical language, we could describe Einstein's work on the so-called theory of special relativity as the discovery that the invariance group enjoyed by electromagnetism is actually $SO(3, 1)$, not merely $SO(3)$: the rotations of 3-dimensional space are supplemented by the rotation of space and time into each other. (Note the comma separating 3 and 1 to indicate that space and time are intrinsically different. Incidentally, as a mark of sophistication, you should say $(3 + 1)$-dimensional spacetime rather than 4-dimensional spacetime.)

Nature sprang a surprise: a vast internal space

The next development was a bit unexpected. By 1920, the proton was recognized as an elementary particle, but twelve years later, surprise, its "uncharged" twin, the neutron, was discovered. Henceforth, we know that atomic nuclei consist of protons and neutrons, generically known as nucleons, as was already mentioned in chapter 4. Under the strong nuclear force, the two nucleons behave as if they were the same particle. The only difference is that the proton carries an electric charge, while the neutron does not, hence its name. Furthermore, their masses are almost exactly the same: the mass of the neutron was measured to be[25] 939.57 MeV, compared to the proton mass of 938.27 MeV. The fractional difference $(939.57 - 938.27)/939.57 = 0.00138$ is less than two parts in a thousand. (In my opinion this "bizarre" fact is not emphasized enough in the popular physics literature.)

Why should Nature throw into the universe a particle that, in almost all important respects, is identical to the proton except that it has no electric charge? I regard this as another example of Her kindness to physicists, giving them a hint of a hitherto unknown symmetry.

It was easy for theorists in the 1930s to imagine that in a universe without electromagnetism, which after all is orders of magnitude more feeble than the strong interaction** (hence the name "strong"), the proton and the neutron would have the same mass. The effects of electromagnetism on subnuclear particles should be tiny. They proposed that, in a world with electromagnetism "switched off," the proton and neutron would have the same mass and behave in exactly the same way under the strong interaction. The two nucleons are then truly the Tweedledee and Tweedledum of the microscopic quantum world: there is no way to tell them apart.††

People are sometimes puzzled by theoretical physicists talking about switching this or that interaction off. Where is the "switch"?

Well, you will learn in chapter 10 that physics is formulated nowadays as an action, which consists of a bunch of terms added together. To switch off electromagnetism, we simply omit those terms that pertain to electromagnetism.

Heisenberg and others proposed that the strong interaction is invariant under the group $SU(2)$, which "rotates" the proton and the neutron into each other.[26] Yes, the 2 in the name $SU(2)$ has to do with the two nucleons being rotated: 2 instead of the 3 directions in the space that we know and love, east and west, north and south, and up and down, rotated into each other by the group $SO(3)$. You would have noticed also that the O has been changed to a U. This is because in everyday rotation we are dealing with real numbers such as the sine and cosine of various angles

---

**We will discuss the four fundamental interactions in more detail in chapter 7.

††I am simplifying here. Nowadays, the mass difference between the proton and the neutron is attributed not only to electromagnetism but also to the mass difference between the up quark and the down quark, the origin of which is not yet understood.

that some schoolchildren learn. In contrast, the transformation of a proton and a neutron into each other corresponds to, roughly speaking, a rotation in a 2-dimensional space constructed out of complex numbers that quantum physics famously traffics in. We mention all this here without going into details at the risk of mystifying the reader even more.[27]

This symmetry, known as isospin, is of great importance in the history of theoretical physics because it opens up a vast internal space. Previously, all the symmetries considered in physics reside in the space and time we were born into, viz the rotation group $SO(3)$ and the Lorentz group $SO(3, 1)$ introduced by Einstein to rotate space and time into each other. In contrast, $SU(2)$ operates in an abstract internal space spanned by the proton and neutron fields (introduced in the preceding chapter.).

The apparently peculiar name "isospin" was abbreviated from isotopic spin. Atomic nuclei with the same number of protons but different number of neutrons were known as isotopes of each other. And the word "spin"? You guessed it, the same group $SU(2)$ was used earlier to deal with the electron spin, as was discussed in chapter 3. The electron spinning up and the electron spinning down are rotated into each other.[‡‡] (Up was metaphorically associated with the proton, down with the neutron, and these historical mental clutches then morphed into the names up quark and down quark, which we will meet shortly. If you feel that the history of physics is rather convoluted, you are right!)

Yet another example of Nature's kindness! Physicists didn't even have to learn more group theory. Mathematicians couldn't care less what physicists were rotating. As long as two "entities" were rotated into linear combinations of each other in the quantum world, then it is fun time with $SU(2)$!

From *SU(2)* to *SU(3)*

In the 1960s, Murray Gell-Mann[28] (known as MGM to some), Yuval Ne'eman, and others[29] enlarged the invariance group of the strong interaction from $SU(2)$ to $SU(3)$. Yes, the 3 has to do with the 3 quarks.[30] As I said earlier, the number of states in a specific representation, namely its dimension, for a group is an immutable property of the universe. For instance,

---

[‡‡]Some readers might wonder how this relates to the actual 3-dimensional rotation in the space we live in. To understand this, physicists had to learn a subtle piece of mathematics: the group $SU(2)$ double covers $SO(3)$. Even almost a hundred years later, some students are still bedeviled by this.

the dimensions of the five smallest representations of $SU(3)$ are, respectively, 1, 3, 6, 8, and 10. These numbers[31] are fixed by mathematicians, or if you prefer, by God, and they are nonnegotiable.

In the necessarily breathless account of this book, we have to reduce enormous advances to a mere sentence or two. MGM proposed that strongly interacting particles, known as hadrons, are composed of quarks. In particular, the proton consists of two up quarks and one down quark, denoted in brief by (uud), the neutron of one up quarks and two down quarks, denoted by (udd). Physicists thus learned that the transformation proposed by Heisenberg amounts to rotating the up and the down quarks into each other.[32]

A spectacular triumph of symmetry considerations and group theory occurred in 1964. At the time, experimentalists had established that there existed nine baryon resonances with roughly the same mass. (These are short lived cousins of the proton and neutron, existing only fleetingly in particle accelerators. For our purposes here, it is not essential to know more about these resonances.) Gell-Mann looked at his list of numbers for $SU(3)$ (which I gave you earlier) and saw that 9 was not allowed. It could only be 10! With supreme confidence, he told the experimentalists to go and find the tenth particle, which he named the Omega minus. Sure enough, when an experimental team led by Nick Samios[33] (see figure 7) turned on their multizillion dollar accelerator and detector, a particle with the predicted properties of the Omega minus popped out. Nobel prize[34] etc.

Symmetry breaking: explicit and spontaneous

Some symmetries of physics are exact (as far as we know), such as rotational invariance. Others are broken explicitly. For example, isospin is broken slightly by the mass difference between the proton and the neutron. From the late 1950s to the early 1960s, it gradually dawned on physicists that symmetries could also be broken spontaneously. Consider a ferromagnet. The laws of physics governing the behavior of the iron atoms inside a ferromagnet manifestly respect rotational invariance, but yet the ferromagnet manages to generate a direction defined by the magnetization, thus breaking the actual symmetry $SO(3)$ down to the group of rotations in 2-dimensional space around this direction, namely $SO(2)$. Now imagine a race of miniature physicists living inside the ferromagnet. All they observe is the symmetry $SO(2)$. Is it possible for them to deduce that the underlying

Figure 7. From left to right: MGM, MGM's lady friend, Nick Samios, and the author.

true symmetry is actually $SO(3)$? The answer is yes, arrived at after much confusion and struggle and rewarded by a Nobel prize for the Japanese American physicist Yoichiro Nambu.

"Symmetry dictates design"

In my 1986 book *Fearful Symmetry* I pushed the view that one of Einstein's legacies to physics is the primary importance of symmetry in leading us into terra incognita.

Take the glory of 19th century physics, namely the understanding of electromagnetism. Starting with lodestones and ember rubbed with fur, physicists labored mightily, experimenting with copper wires, cardboard stacked and soaked with acid, frogs' legs, and the like, They gradually distilled a multitude of observations into several equations, which Maxwell synthesized into an organic whole. But the underlying symmetry remained hidden and unrecognized until the likes of Lorentz and Einstein came along. Electromagnetism was found to be invariant under the Lorentz transformation (which we will discuss in detail in chapter 6) described by the group $SO(3, 1)$ (as was already alluded to above). A symmetry of spacetime was discovered.

But the mathematical steps could also be run in the other direction. We could have imposed Lorentz invariance on the electric attraction between two charges and thus obtain the full set of Maxwell's equations. Symmetry dictates the electromagnetic interaction.

Einstein pioneered this way of looking at physics. In particular, he realized that Lorentz invariance, being a symmetry of spacetime, necessarily must also be imposed on mechanics. The celebrated formula $E = mc^2$ (which we will also discuss in some detail in chapter 6) pops out as a consequence.

Theoretical physics is a vast subject that, roughly speaking, could be divided into two. In many areas, the underlying laws are known, and the challenge is to extract and understand the consequences of those laws. Fluid dynamics amounts to applying Newton's laws of motion to fluids. The laws have been known for centuries, but there is still little hope of deriving turbulence from these laws. Condensed matter physics involves the electromagnetic attraction between electrons and ions in various arrangements, governed by quantum laws. But it was far from simple to derive the astonishing phenomenon of superconductivity from first principles; more than half a century of challenging work was needed.

In contrast, in other areas, notably particle physics from around 1930 till the 1980s, the underlying laws were totally unknown. Then Einstein's precept of "symmetry dictates design" comes to the fore. Thus, the postulated symmetry of isospin constrains the possible structure of the strong interaction, about which practically nothing was known in 1930. The story continues with Gell-Mann's extension of isospin, from the group $SU(2)$ to $SU(3)$ as was noted earlier. Eventually, the law governing the strong interaction was discovered, as will be told in chapter 7. After that, particle physics moves into the other arena, of physics concerned with extracting the consequences of known laws.

I have known ever since I started studying physics that Einstein was largely responsible for locating symmetry at the core of theoretical physics, but still I was astonished recently at the extent to which symmetry has risen in the hearts and minds of physicists. I read that aside from discussions of crystals, the word "symmetry did not appear"[35] in the discourse of physics until surprisingly late. For instance, the definitive encyclopedia of physics of 1929 had only one reference to symmetry, and that is an essentially irrelevant reference to the interchange symmetry of the metric tensor appearing in Einstein's theory of gravity (which we will describe in chapter 6).

"Spiritual formulas" in physics?

You have no doubt noticed that two themes run through the chapter, conservation laws and symmetry, but so far independently of each other. I started this chapter by recalling that, when you and I first learned about conservation laws in physics, our teachers or professors were unable to tell us where they came from. These laws were presented as empirical facts. Of course, since physics is an empirical science, any statements in physics, let alone laws, should be experimentally verified. What we would like to have is a deeper understanding.

Having gotten this far into this chapter, you naturally suspect that conservation laws and symmetry are connected. Indeed, conservation laws originate in the symmetries of physics.

The profound connection between them was discovered by Emmy Noether in what is now known as Noether's theorem.[36] (See figure 8.) I am here to tell you how important this connection is, but I can't do it any better than the grand old man of physics, Albert himself!

> Pure mathematics is, in its way, the poetry of logical ideas. One seeks the most general ideas of operation which will bring together in simple, logical and unified form the largest possible circle of formal relationships. In this effort toward logical beauty spiritual formulas are discovered necessary for the deeper penetration into the laws of nature.
>
> —Albert Einstein, writing about Amalie Emmy Noether

I believe that, in this memorial statement, by "pure mathematics" Einstein meant theoretical physics, in line with the usage in his time.

In our age of enlightened diversity, an abundance of popular articles has popped up extolling women in physics. These articles invariably trot out a parade of the "usual suspects" headed by Madame Curie. Absolutely no question that Marie Curie is exceedingly first rate among physicists, male or female. But to my amazement, almost none of these articles ever mentions Emmy Noether, and no doubt many of their authors have never heard of her. I dare say that none of the theorists exalted in the media has come anywhere close to discovering a formula Einstein would judge so

Figure 8. Emmy Noether, probably the most famous woman physicist in theoretical physics whom almost nobody outside of physics has ever heard of. Photographer unknown. Image courtesy of Dr. Monica Noether (CC BY-ND 2.0).

beautiful as to call it "spiritual." What Noether discovered is something truly amazing, something fundamental to the universe, not one of those amazing discoveries reported breathlessly on the web, here today and gone tomorrow.

So, why is energy conserved? Noether tells us that it is because the laws of physics are the same yesterday, today, and tomorrow. Speaking in jargon, physicists say that energy is conserved because of time translation invariance. Or, in plain English, the law does not change with time.[37] Remarkably, we could begin to glimpse the connection between energy conservation and time translation invariance via a simple contrarian example. In Newtonian mechanics, when a ball is kicked, the law governing its motion changes for that instant, and the ball's energy is for sure not conserved.

Similarly, Noether answers the question of why momentum is conserved: it is because the laws of physics are translation invariant in space. Again, consider the contrarian example of a ball rolling down a hill. The law governing the ball's motion changes depending where it is on the hill, and so momentum is no longer conserved.

These two examples are so simple, almost laughably so, that they risk demeaning the full power of Noether's theorem. We will see in chapter 10 that in modern physics the laws are expressed not in simple equations such as Newton's $F = ma$, but in terms of a quantity known as the action. When theoretical physicists are presented with an unfamiliar action, they immediately examine it for the symmetries it exhibits, and invoke Noether's theorem to deduce the conservation laws this action would lead to. Nowadays, Noether's theorem is used almost subconsciously.

Here is a table illustrating the intimate relationship between conservation laws and symmetries.

conservation laws	symmetries
energy conservation	invariance under translation in time
momentum conservation	invariance under translation in space
angular momentum conservation	invariance under rotation
charge conservation	invariance under multiplication of charged fields by a complex phase

Deeper and simpler: the "poetry of logical ideas"

As physicists explore Nature at ever deeper levels, Nature appears to get ever simpler, at least thus far, as I have said repeatedly. Surely a stunning truth about the universe we inhabit, and a gift physicists may or may not deserve!

To understand this, let us return to the psychiatrist we encountered earlier in this chapter. Suppose she is actually studying human memory. A few days after showing you a circle and the irregularly shaped polygon shown in figure 2, she asks you to describe these two shapes. Recalling an "irregularly shaped polygon" does not suffice. You must answer questions such as how many vertices does the polygon has? What are roughly the angles at each vertex? And so on.

Of course, the circle would be much easier to remember than this crazy polygon.

Our silly example provides a pale and merely metaphorical representation of the contrast between elementary and advanced physics. Consider mechanics. We know that at a deeper level Einsteinian mechanics rules, and Newtonian mechanics is but an approximation to Einsteinian mechanics. I will discuss Einsteinian mechanics in some detail in the next chapter, but for now you only have to know that in Einsteinian mechanics the speed of light c imposes a speed limit on the motion of material particles. In everyday life, the typical speed v of material particles is far smaller than c. In other words, c is effectively infinite compared to the speeds the lumbering giants that we could normally attain. The ratio v/c may be set equal to 0 to an excellent approximation. When you do this, Einsteinian mechanics reduces to Newtonian mechanics.

An elementary fact of arithmetic is that when your multiply a small number by another small number you get an even smaller number: for example $(0.1)^2 = 0.01$, $(0.1)^3 = (0.1)^2 \times 0.1 = 0.001$, and so on. We are invited to expand the formulas in Einsteinian mechanics as a series in powers of (v/c), that is, as a sum in which all terms proportional to $(v/c)^k$ with $k = 0,\ 1,\ 2, \cdots$ are collected together. For v much less than c, the terms proportional to (v/c) are small, but the terms proportional to $(v/c)^2$ are even smaller, the terms proportional to $(v/c)^3$ are yet smaller, and so on.

As I have just said, if we simply set v/c to 0 we recover the familiar Newtonian mechanics. But we could easily do better: if we keep only the terms proportional to v/c, then we obtain the first relativistic correction to Newtonian mechanics. Next, keep also the terms proportional to $(v/c)^2$. You could keep on going till you drop of exhaustion. This is known as the post-Newtonian approximation, largely unnecessary for everyday purposes and even for much of physics.

Now let me ask you. Why should Nature care that we are unable to propel ourselves up to speeds comparable to the speed of light? Why should Nature care whether the results of an approximation imposed by humans look pretty?

In Newtonian mechanics, space and time are treated on different footing. With the passage of time, particles move through space. Energy and momentum are manifestly different concepts. For one thing, momentum is directional, but energy is not. In contrast, in Einsteinian mechanics, space and time are unified into a single spacetime through which we and all particles voyage. Energy and momentum are also unified. When we go from Newtonian mechanics to Einsteinian mechanics, the symmetry group $SO(3)$ is extended to $SO(3, 1)$, as has already been mentioned in this chapter.

Einsteinian mechanics is much more symmetrical than Newtonian mechanics, sort of like the circle compared to the irregular polygon in the psychiatric test. Consequently, to educated eyes, Einsteinian mechanics is so much easier than Newtonian mechanics. A short while ago, I would have surprised many readers by saying that Einsteinian mechanics, once mastered, is intrinsically simpler than Newtonian mechanics. But now you have a least an inkling why.

To say the contrary, that Newtonian mechanics is easier than Einsteinian mechanics, is akin to saying that Americans find English easier to understand than Swahili. Of course, we all "understand" Newtonian mechanics better than Einsteinian mechanics. But that is merely because almost immediately after birth we have to grapple with Newtonian mechanics. To the trained eyes of a theoretical physicist in my circle, a formula in an undergraduate textbook treating space and time on different footings might look as awkward and bizarre as a man wearing a boot on one foot and a sandal on the other.

As another example, theoretical physicists in my circle grew up in the relativistic and quantum world and can deal with quantum electrodynamics with ease, referring to it as a fairly simple, almost trivial gauge[§§] theory. They could calculate Feynman diagrams practically blindfolded. But if you give these theoretical physicists an exam on undergraduate level electromagnetism, they might fail miserably unless they were actually teaching the course.[38] That's because at the undergraduate level, the electromagnetic field, born an organic whole in the relativistic world, is forcibly torn apart into an electric field and a magnetic field. The underlying symmetry $SO(3, 1)$ is all but hidden.[39]

I like to say that electromagnetism is native to the relativistic world, where the Lorentz invariance of the theory is manifest. Recognizing the relativistic invariance of electromagnetism, fundamental physicists now write Maxwell's equations more compactly as one equation. When I was a student, I had to memorize Maxwell's equations before every examination.

_____

[§§] How a word normally associated with railroad tracks and wine casks became a megastar in physics due to an erroneous understanding will be explained in chapter 7.

Mmm, let's see, a magnetic field changing in time produces an electric field changing in space—or, is it changing in time? With relativistic invariance, a single equation describes an electromagnetic field changing in spacetime. I find this completely symmetrical equation as easy to remember as the shape of the circle.

Intrinsically, advanced physics is simpler than elementary physics—a little secret not often revealed to the layperson. Many people are stumped by high school or college physics because they are presented with mis-shapen phenomenological equations having little to do with Nature's intrinsic essence, with Her beauty.

A common observation, strange but true: in graduate school, many students of pure math, particularly those from colleges and universities far from the cutting edge of world class research and who may be deluded into thinking that acing an undergraduate course in, say, partial differential equations meant that they could be a leading mathematician, are blown away by the gale force winds of abstraction, layers upon layers of mental intrigues with nary a concrete example in sight. In sharp contrast, in theoretical physics some students, as if struck by some divine light, might feel a sudden revelation of how all that messy undergraduate stuff comes together into a symmetrical whole.

In short, graduate school is where many students in pure math discover how difficult their chosen subject could get, while students in theoretical physics are stunned, and delighted, by how easy physics really is. At least that was how I felt.

Notes

[1] More on kinetic and potential energy in chapter 10.

[2] See, for example, *GNut*, pages 29–30.

[3] Andrew Szanton, *The Recollections of Eugene P. Wigner: As Told to Andrew Szanton*, Basic Books, 2003.

[4] Some readers might realize that this rotation could also be redescribed as a rotation though $-120°$.

[5] The invariance group of an N-sided regular polygon has $2N$ elements and is known as the dihedral group D_N. See *Group Nut*, page 60. Hurray for the jargon guy!

[6] Again, forgive me for blurring over various technicalities. A group could also be infinite but discrete. For details, see my book *Group Nut*.

[7] Mathematicians say "composing transformations."

[8] Yup, you got it: rotations in N-dimensional space form the group $SO(N)$.

[9] See the figure on page 124 of *Fearful*.

[10] If you ever encounter a mathematician, you could tell her the following painfully nerdy joke. What do you call a purple fruit who commutes to work? An abelian grape.

[11] Szanton, *The Recollections of Eugene P. Wigner*, page 107.

[12]When I wrote my trilogy of textbooks, I decided to write about what I considered the most beautiful subjects in theoretical physics, in this order: quantum field theory, Einstein gravity, and group theory.

[13]No question that these discrete symmetries furnish one of the most fascinating chapters of modern physics! I refer the interested readers to my books *Fearful* and *QFT ASAP*.

[14]We will encounter antimatter in chapter 6.

[15]They sure could tell if allowed to invoke biology.

[16] For a short article on time reversal, I suggest my essay in *Mysteries of Life and the Universe: New Essays from America's Finest Writers on Science*, edited by W. H. Shore, Harcourt, 1992.

[17]One to specify the angle of rotation, and two to specify the direction of the rotational axis.

[18]Due to the uncertainty principle, which forbids us to determine position and momentum at the same time, we have to abandon the concept of an orbit.

[19]For those readers wondering why they are not called ψ_5, ψ_4, ψ_3, ψ_2, ψ_1. Good question! But the answer is not something we need to go into for the purpose of this book. Those readers who want to learn about this peculiar nomenclature should consult a book on modern physics, such as the book by Leighton cited in the figure caption.

[20]That it is a sum is mandated by the postulates of quantum physics.

[21]Some readers would know that I exaggerated somewhat when I spoke of the irrelevance of group theory for classical physics. The important thing is the ability to superpose or form linear combinations of the quantities we are dealing with. We cannot superpose orbits, but we could superpose electromagnetic fields. That was why I picked on Newton, but not Maxwell.

[22]Thus, the applicability of group theory to physics depends crucially on whether the relevant physics is linear or nonlinear. Quantum physics is linear, while classical physics is nonlinear. Hence the story of the bragging mathematician in the 19th century.

[23]For the sake of completeness, I mention in passing the "opposite" phenomenon of emergent or apparent symmetry. The materials studied by condensed matter physicists are governed ultimately by the electromagnetic interaction between electrons and atomic nuclei, whose symmetries have been known for more than a century. Yet when these materials are looked at through "blurry glasses," they could effectively exhibit symmetries that are not there at the fundamental level.

[24]Several of my books are devoted to these symmetry groups: *Unity, Fearful, Group Nut, QFT ASAP*.

[25]I won't bother to take time to explain what the mass unit MeV means; it is not relevant to the story being told. I also won't bother to include the error bars.

[26]Heisenberg originally had merely an interchange symmetry Z_2, namely the invariance group of the isosceles triangle mentioned more than once earlier. The full $SU(2)$ group came later and was due to B. Cassen and E. U. Condon, *Phys. Rev.* 50 (1936), 846, and G. Breit and E. Feenberg, *Phys. Rev.* 50 (1936), 850.

[27]For those who would like to know, O stands for orthogonal, U for unitary. Look at a group theory text, such as *Group Nut*, and all will be clear. Just to calibrate in case you want to learn more: most of the undergraduates in my group theory course have no trouble grasping all this stuff.

[28]For an insightful biography of this giant who dominated particle theory for decades, see G. Johnson, *Strange Beauty*, Vintage, 1999.

[29]The earliest effort was due to the Japanese physicist Shoichi Sakata.

[30]Named up, down, and strange by Gell-Mann. Quarks actually occurred to MGM after $SU(3)$.

[31]Read a group theory textbook, such as my *Group Nut*, and you will be able to deduce these mysterious numbers. Yes, they all stem from the number 3.

[32] In order to make the electric charge of the proton come out to be $+1$ and that of the neutron come out to be 0, MGM had to make the up quark and the down quark carry charge $+\frac{2}{3}$ and $-\frac{1}{3}$ respectively, as you can see for yourself by solving a simple algebra problem. (Since the proton consists of (uud), the neutron of (udd), the quark charges have to satisfy $2Q_u + Q_d = 1$, $Q_u + 2Q_d = 0$. You can easily solve these two equations to find $Q_u = +\frac{2}{3}$ and $Q_d = -\frac{1}{3}$.) Up to that point, physicists had never encountered fractional charges before, and so many objected. In hindsight, it seems a bit silly.

[33] For a first person account, see N. Samios, pages 31–39 in *Proceedings of the Conference in Honour of Murray Gell-Mann's 80th Birthday*, edited by H. Fritzsch and K. K. Phua, World Scientific, 2011.

[34] Among physicists at that time, a Nobel for Gell-Mann was such a foregone conclusion that for years he would eagerly await the announcement from Sweden. In those days before easy transcontinental phone calls (let alone emails), my PhD advisor Sidney Coleman once traveled to Stockholm in the late fall and sent Gell-Mann a telegram with only the words "Ignore previous telegram" just to drive MGM crazy.

[35] G. Holton, *The American Scholar*, 41 (1971–1972), page 101. I thank P. Galison for alerting me to this paper.

[36] For some examples of Noether's theorem in action, see chapter II.4 of *GNut*.

[37] The word "translate" has the same root as the word "transport," in the sense of carrying from one language to another. In physics, it is used in the sense of transport from the past into the future, or from one region of space to another. In Greece, moving vans often have the word "translate" painted on their sides, in the Greek alphabet of course.

[38] Many of my friends in physics say that if you want to understand a subject, the best way is to teach a course about it. Murph Goldberger, the chair of the Princeton physics department at the time of this story, once in desperation asked me to teach the graduate course on many body physics. The students had submitted a petition asking that the professor who taught the course not be allowed to teach it again. When I protested that I knew literally nothing about the subject, he answered that since he believed that I was more intelligent than the average graduate student enrolled in the course I would do fine if I simply read the textbook before the students did. Later in life, I felt grateful that he had forced me to learn many body physics and to learn it enough to tell a roomful of students about it.

[39] As an example, contrast the expressions for the power radiated by a charged particle given to graduate students and to undergraduate students, respectively (3) and (4) on pages 79 and 80 of *FbN*. What is fed to undergraduates is all misshapen, with the underlying Lorentz invariance mashed up.

6
EINSTEIN, THE EXTERMINATOR OF RELATIVITY AND THE CHOREOGRAPHER OF SPACETIME

Do you know who invented the "theory of relativity"?

Einstein had two daughters, scintillating but rather unhappily named special and general relativity. Later in life, Einstein regretted calling his work the theory of relativity.

Regret, yes. Calling it special relativity, no!

You could search Einstein's epoch-making 1905 paper[1] for the phrase "theory of relativity." You won't find it! This unfortunate term, the theory of relativity, was coined by a totally unknown German physicist named Alfred Bucherer, long lost in the dustbin of history and possibly an anti-semite. Bucherer was the first to use, in 1906, the term[2] "Einsteinian relativity theory" while criticizing Einstein's theory.

So, the answer to the bet-winning "barroom question" is that Bucherer invented the term "theory of relativity," but not the theory.

I must now vent my pet peeve. Physics contains a number of unfortunate names, some due to historical confusion long since cleared up. Probably the worst name ever is relativity, as it has spawned a swarm of nonsensical statements, such as "Physicists have proved that truth is relative" and "There is no absolute truth; Einstein told us so," uttered with smug authority by numerous ignorant fools. It is only a slight exaggeration to say that Bucherer had inadvertently messed up the minds of more than a few "eminent" philosophy professors.

Yes, of course physicists talk about observers moving relative to each other—that is part of everyday language—and hence unavoidably use the word relative, but what they search for are the invariants, meaning those quantities that do not vary, that is, remain unchanged. Physicists search for quantities different observers could agree on, and require that the laws of physics be invariant and independent of observers.

Einstein promoted the exact opposite of "truth is relative," and once said that he should have named his work invariant theory.

Galilean relativity

Unfortunately, the word "relativity" is by now so encrusted in, and beyond, physics that there is no way to change it. I so wish that Einstein had called it "invariant theory"! I, and many others, think that two better names for Einstein's theory of special relativity and theory of general relativity are spacetime physics and Einstein gravity[3] respectively.

To understand Einstein, we must first understand Galileo, but there is nothing to understand, because what Galileo said is just ordinary common sense. Everybody understands the everyday notion of how velocities add. Consider a person walking briskly at 1 meters per second on an airport conveyor belt moving by at 2 meters per second. The guy in a souvenir shop watching the passengers go by sees this person moving by at $1 + 2 = 3$ meters per second. Of course. And if the person on the conveyor belt is standing still, the guy in the souvenir shop would see this person moving by at $0 + 2 = 2$ meters per second.

That velocities add is known to physicists as the Galilean addition of velocities, but you need hardly to have taken a physics course, nor to have even heard of Galileo, to understand this. But inexplicably, in an introductory course, some students are already trembling at this point with fear and loathing.

Let's formalize this a bit and define an inertial frame of reference as one attached to an observer moving smoothly in a straight line at a fixed velocity, such as the traveler on the conveyor belt or Galileo in his sailing ship. Update Galileo's ship to Einstein's train.[4] The passenger on the train, an acquaintance of ours named Ms. Unprime, ascribed to some event the spatial coordinates (x, y, z) and temporal coordinate t. To the same event, the observer on the ground, Mr. Prime, assigns the coordinates (x', y', z') and t'. See figure 1.

Unavoidably, this chapter will contain a touch of mathematics. But fear not, merely high school level algebra and the teeniest bit of differential calculus.

Denote the speed of the train by u and choose the axis so that the train moves along the x-axis. Then the two sets of coordinates are related by

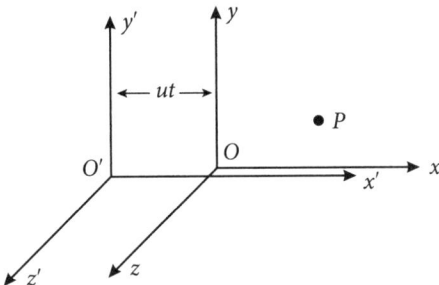

Figure 1. Two inertial frames moving with velocity u relative to each other. At time $t = t'$, a point P with coordinates (x, y, z) in one frame of reference would evidently have coordinates $(x', y', z') = (x + ut, y, z)$ in the other frame of reference.

$$t' = t, \quad x' = x + ut$$

(To keep the algebra simple, we have set the clocks and chosen the origins of the two inertial frames, $x = 0$ and $x' = 0$, to coincide at $t = 0$ and $t' = 0$.)

The two coordinates perpendicular to the direction of motion do not[5] transform: in other words, $y' = y$ and $z' = z$. They are just going along for the ride. To avoid clutter, we will keep them on the back burners of our minds instead of writing them out explicitly. These relations between (t, x, y, z) and (t', x', y', z') are known as the Galilean transformation. The equation about time, $t = t'$, states that time is universal, ticking away at the same rate for different observers moving relative to each other at a constant velocity.

The equation about space, $x' = x + ut$, merely defines the relative velocity u. Consider a point (Ms. Unprime's seat for example) on the train with $x = 0$. Plugging this into this commonsense equation, we see that, to Mr. Prime on the ground, this point moves along according to $x' = ut = ut'$. She is indeed moving with velocity u in the x direction.

Note that $x' = x + ut$ implies $x = x' + (-u)t'$. To Ms. U, she is at rest, while Mr. P, standing at $x' = 0$, is moving relative to her with speed $-u$ in the x direction.

Addition of velocities

The Galilean law of velocity addition follows almost immediately. Suppose Ms. U tosses a ball forward with velocity* $v = x/t$, what is the velocity of

---

*Beginning students are often warned to distinguish between velocity and speed, but for convenience I will use the two words interchangeably.

the ball as seen by Mr. P, namely $v' = x'/t'$? By definition, it is equal to distance traveled divided by time elapsed, and thus

$$v' = \frac{x'}{t'} = \frac{x + ut}{t} = \frac{x}{t} + u = v + u$$

Velocities add, exactly as common sense tells us. The algebra is hardly necessary.

Of course, v' is not equal to v. The two observers in motion relative to each other should not, would not, and could not possibly agree on how fast they measure the ball to be moving. Even a child understands this. Does this mean "Truth is relative"? Really, give us a break.

The laws of physics must not depend on the physicist

The velocity of an object as measured by two observers moving relative to each other are different for sure. In contrast, the laws of physics governing the motion of the object as deduced by the two observers must be absolutely the same.

It is an extremely simple truth: physics must not depend on the physicist. Indeed, Galileo already understood this basic point when he talked about candle flames moving straight up and butterflies flying around in a cabin on a smoothly sailing ship. Not to mention that you are able to pour your favorite drink on a smoothly flying airplane without spilling.

Remember this: Einstein exterminated the relativity of truth.[6] Yes, put a cape on that Einstein and let him fly around and exterminate those who knoweth not what they speaketh.[7]

Wresting time and space back into physics

You might have thought that I was making up what I said about philosophers, but a distinguished philosophy professor once asked me to explain Einstein's theory of relativity, and by way of bragging, he added that he did well in his high school calculus class. After I got through the Galilean addition of velocities, he beamed, complimented me for giving such a clear explanation, and went away happily ignoring my attempt to tell him this was all 17th century stuff and that we hadn't gotten to Einstein yet. Needless to say, I have also met much better informed philosophers since then. In particular, as I was writing this chapter, a philosophy professor flew across the United States to talk to me about one of my popular books. Over lunch, he told me that Feynman and I, though never missing an opportunity to mock philosophers, are "actually quite philosophical in your individually peculiar ways." I was of course delighted to hear this, and I had to admit that we are all united in seeking the truth.[8]

> I am convinced that the philosophers have had a harmful effect upon the progress of scientific thinking in removing certain fundamental concepts from the domain of empiricism, where they are under our control, to the intangible heights of the a priori. . . . This is particularly true of our concepts of time and space, which physicists have been obliged by the facts to bring down from the Olympus of the a priori in order to adjust them and put them in a serviceable condition.[9]

I understand that Einstein was berating philosophers for yet another sin, throwing words around without defining them operationally.

"I am convinced that the philosophers have had a harmful effect upon the progress of scientific thinking. . . ." Well, I didn't say it, but I certainly couldn't agree more.

How to allow light to evade Galilean relativity?

Historically, Einstein's deep insight is that this simple everyday common-sense addition of velocities must fail for light. Maxwell told us, as the crowning glory of 19th century physics, that light is an electromagnetic wave moving along, alternating between an electric field and a magnetic field, as was discussed in chapter 4. An electric field changing in space and time generates a magnetic field changing in space and time, which in turn generates an electric field changing in space and time. So on and so forth: an electric field and a magnetic field ceaselessly changing into each other. And voilà, a wave! Thus, the speed of light, denoted by c as you surely know, in honor of the Latin word *celeritas*, is fixed by how varying electric and magnetic fields generate each other, and does not depend on who, Ms. Unprime or Mr. Prime, shines the beam of light forward. But this manifestly contradicts the Galilean addition of velocities: $c \neq c + u$!

Starting with this fact, that the speed of light c is a universal constant, Einstein derived his invariant theory with a few lines of high school algebra, and showed that c imposes the ultimate speed limit. The Galilean addition of velocities, which still must hold for speeds small compared to c, is replaced by what is known as the Lorentzian addition of velocities. The question is how to modify $t' = t$ and $x' = x + ut$ so as to evade Galilean addition of velocities for light.

I stated in the prologue that I envisage a range of readers. The rest of this chapter involves a bit of high school math. Some readers may want to skim over this material, but I could promise those readers who read attentively and check the equations a significant reward.

Einstein's clock

> I remember that an uncle told me the Pythagoras theorem before the holy geometry booklet had come into my hands. After much effort I succeeded in "proving" this theorem on the basis of the similarity of triangles: in doing so it seemed to me "evident" that the relations of the sides of the right-angled triangle would have to be completely determined by one of the acute angles.
>
> —Einstein speaking of one of his two[10]
> formative childhood experiences[11]

If you like, you could say that Einstein generalized the schoolchild's understanding of space to space and time.

Many roads lead to the Lorentz transformation. I choose to give you Einstein's original argument. Consider a simple clock consisting of two mirrors, separated by the distance L, between which a pulse of light bounces back and forth. See figure 2(a). When the clock ticks, the pulse leaves the bottom mirror, reflects off the top mirror, and comes back to the bottom mirror, causing the clock to tock.

Let Ms. Unprime hold this clock oriented along the y-axis in her frame. To her, the time between tick and tock is just $T = 2L/c$.

But to Mr. Prime on the ground, the clock tocks at time T' which may well differ from T. Let us determine T' using high school math; I will go extra slow to make sure you follow. By the time the clock ticks, the bottom mirror has moved a distance of uT' and the pulse of light has traveled a distance of cT'.

To determine T', let us calculate the distance the pulse of light covered by using good old geometry. We see in figure 2(b) that two right-angled triangles stacked back-to-back are involved. Focus on one of the triangles. Its height is simply L, while its base is half the distance the mirror has moved, namely $\frac{1}{2}uT'$. Pythagoras tells us that the hypotenuse has length $\sqrt{L^2 + (\frac{1}{2}uT')^2}$. We have to double this to obtain the distance the pulse of light covered, as seen by Mr. Prime. Thus, this distance equals $2\sqrt{L^2 + (\frac{1}{2}uT')^2}$. But we also know that light has traveled a distance

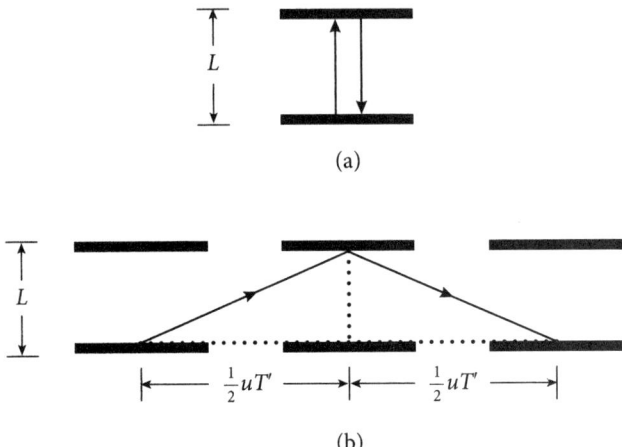

(a)

(b)

Figure 2. Einstein's clock as seen by the two inertial observers, Ms. Unprime and Mr. Prime. (a) To Ms. Unprime, the clock is not moving, and light takes time $T = 2L/c$ to traverse the round trip distance of $2L$. (b) To Mr. Prime, however, by the time the clock tocks it has moved forward by a distance of uT', just by the definition of u and T'. To determine the distance light has traveled according to Mr. Prime, Einstein asked Pythagoras for help. We have two right-angled triangles formed by the path light traveled and the dotted lines drawn to guide the eyes connecting the two mirrors at various instants in time. The two dotted sides of each right-angled triangle have lengths L and $\frac{1}{2}uT'$ respectively, and thus the hypotenuse has length $\sqrt{L^2 + (\frac{1}{2}uT')^2}$. The distance light has traveled by the time the clock ticks equals twice the hypotenuse of each of these right-angled triangles, namely $2\sqrt{L^2 + (\frac{1}{2}uT')^2}$.

of cT'. Equating these two quantities $cT' = 2\sqrt{L^2 + (\frac{1}{2}uT')^2}$ gives us an equation to determine T'.

But due to a clever argument by Einstein, we don't even have to solve for T'. Square both sides of this equation: the left hand side becomes $c^2T'^2$ while the right hand side becomes

$$4\left(L^2 + \frac{1}{4}u^2T'^2\right) = 4L^2 + u^2T'^2 = 4L^2 + X'^2$$

The first equality follows by multiplying through by 4 and using $4 \times \frac{1}{4} = 1$, and the second equality is just our recognition that, by definition of u, the product uT' equals X', the position of Einstein's clock as seen by Mr. P when it tocks. Thus, we obtain $c^2T'^2 = 4L^2 + X'^2$. Subtracting X'^2 from both sides, we find

$$c^2T'^2 - X'^2 = 4L^2 = c^2T^2 - X^2$$

The last equality follows from our earlier observation that $T = 2L/c$ and that $X = 0$ since the clock has not moved in Ms. Unprime's frame. This tells us even though the two observers do not see the same values for space and time, they agree on the combination $c^2 t^2 - x^2$.

To repeat, (T', X') and (T, X) are different, but yet the combinations $c^2 T'^2 - X'^2$ and $c^2 T^2 - X^2$ are the same.

Let us now observe that in this invariant combination the relative velocity u does not appear! Thus, a third observer going by on a train moving at some other velocity relative to Mr. Prime would measure the same value for $c^2 t^2 - x^2$. Indeed, a fourth observer, a fifth observer, ... In other words, all observers in relative motion would agree on the value for $c^2 t^2 - x^2$. This combination is what Einstein, and mathematicians, call an invariant.

As was emphasized earlier in this chapter, physicists love invariants, quantities that are the same for all observers. It is only a slight exaggeration to say that physicists are searching, constantly and all over the place, for invariants. This is almost the exact opposite of what the "truth is relative" crowd claims.

The Lorentz transformation

> He meant more than all the others I have met on life's journey.
> —Einstein speaking of Lorentz

Armed with this invariant, we are ready to modify Galileo's transformation $t' = t$, $x' = x + ut$ to the Lorentz transformation, so that $c^2 t'^2 - x'^2 = c^2 t^2 - x^2$. Remarkably, finding the Lorentz transformation requires yet again no more than high school algebra.

To lessen clutter, let us use units in which $c = 1$. Better than using an absurd unit such as fathoms per week! Inexplicably, some students tense up at this simplifying step. Actually, this is familiar to almost anybody who reads the newspaper or watches movies about interstellar warfare. If we use light year as a unit of distance, that is, the distance that light travels in a year, then the speed of light is just one year per year, namely 1. As you will see, we could always put c back in our final result if we want to.

Write $t' = p(t + qx)$, $x' = r(x + ut)$, where p, q, and r are three unknowns to be determined. Note that x' is still proportional to $x + ut$ because that defines u: to Ms. Unprime the passenger on the train, the point $x' = 0$ where Mr. Prime the station master is standing is moving according to $x + ut = 0$, that is, $x = (-u)t$.

To determine the three unknowns p, q, and r, we impose the requirement that $t'^2 - x'^2 = t^2 - x^2$. A few steps of algebra[12] then lead us to the celebrated Lorentz transformation,[13] in units with $c = 1$:

$$t' = \frac{t + ux}{\sqrt{1 - u^2}} \quad \text{(temporal)}, \qquad x' = \frac{x + ut}{\sqrt{1 - u^2}} \quad \text{(spatial)}$$

Purely for future reference, I have labeled these two equations as "temporal" and "spatial."

Even after so many decades, it still blows my mind to see the deepest secret of space and time revealed by a few lines of high school algebra.[14]

Putting c back

As promised, we could now put c back in if you so desire. Consider the quantity $(1 - u^2)$ inside the square root. If we blindly plug in u equal to some number of meters per second, it makes no sense since you cannot subtract meters squared per second squared from the number 1. (We are of course in no way tied to such humanmade units as meters per second; I use these familiar units just to be definite.) Hence, u being a velocity must be divided by something with dimension of velocity, and the only "something" we have at our disposal is c. Thus, when we restore c, the quantity inside the square root must become $(1 - \frac{u^2}{c^2})$.

In physics and in engineering, this self-evident requirement that units must match is known as dimensional analysis.[15] A deep truth is that only three dimensions[†] are needed for all of physics, namely length L, time T, and mass M. (The letters L, T, M are used to denote length, time, and mass generically in dimensional analysis, and do not refer to any specific quantities.) I use a square bracket to denote the dimension of a physical quantity X as $[X]$. Thus, $[u] = L/T$: velocity has dimension of length divided by time.

Next, the quantity $(t + ux)$ that appears in the Lorentz transformation also makes no sense because $[t] = T$ while $[ux] = (L/T)L = L^2/T$; you can't subtract miles squared per hour from an hour. Thus, to add ux to t, we must divide ux by something with dimension L^2/T^2. Hence, by dimensional analysis, we replace $(t + ux)$ by $(t + \frac{ux}{c^2})$.

[†]Not to be confused with the dimension of space, namely 3, or of spacetime, namely $3 + 1 = 4$!

Behold, the Lorentz transformation, with c restored,

$$ct' = \frac{ct + \frac{u}{c}x}{\sqrt{1 - \frac{u^2}{c^2}}} \quad \text{(temporal)}, \qquad x' = \frac{x + \frac{u}{c}ct}{\sqrt{1 - \frac{u^2}{c^2}}} \quad \text{(spatial)}$$

and, as before, $y' = y$, $z' = z$.

Did you notice that I wrote the temporal equation in terms of ct rather than t? So another way of looking at c: the speed of light is just a conversion factor between space and time.

In the Newtonian world, c is effectively infinite, and so $\frac{u}{c}$ tends to zero. The relation between t' and t becomes simply $t' = t$. Newton's notion of absolute time emerges.

The astonishing world of Einstein's spacetime

Entire books can be written, and many were, on Einstein's theory of special relativity, or spacetime physics as I prefer to call it. Here we content ourselves with a few remarks, epoch changing though they are.

(A) For u much less than c, $(1 - \frac{u^2}{c^2})$ approaches 1 and so the square root disappears. We see that the Lorentz transformation reduces to the Galilean transformation. In this connection, I get somewhat upset whenever I see statements such as "Einstein overthrew Newton." No overthrowing is involved: Newtonian mechanics continues to hold in any situation in which the objects involved are moving slowly compared to c. In physics, a proposed new theory must reduce smoothly to an older and empirically established theory. This poses an often stringent demand on any crazy theory people propose on the internet three times a day before breakfast. There is apparently no such continuity requirement on theories in, for instance, economics.

(B) The speed limit of the universe is set by c. If the velocity u exceeds c, the quantity $(1 - \frac{u^2}{c^2})$ becomes negative, and the square root goes imaginary. The whole thing goes haywire.

(C) We could now easily derive the new law for the addition of velocities. We simply divide the second equation in the Lorentz transformation by the first, thus obtaining the velocity $v' = x'/t'$ seen by Mr. P when Ms. U says the object is moving with velocity $v = x/t$:

$$v' = \frac{x'}{t'} = \frac{x + \frac{u}{c}ct}{t + \frac{u}{c^2}x} = \frac{\frac{x}{t} + \frac{u}{c}c}{1 + \frac{u}{c^2}\frac{x}{t}} = \frac{v + u}{1 + \frac{uv}{c^2}}$$

(For the third equality we divided numerator and denominator by t.) Note that the square roots in the Lorentz transformation cancel out. When the velocities involved are small compared to c, the denominator $1 + \frac{uv}{c^2}$ is approximately 1 and this reduces to the Galilean law $v' = v + u$ for the addition of velocities.

This simple algebraic expression has the remarkable property[16] that if $v = c$, then $v' = c$ also, not $v' = v + c$!

This peculiar law of addition has been verified countless times at particle accelerators. No matter how much energy you pump into a subatomic particle, trying to make it go faster, its speed could approach c as closely as you like, but never exceeds c.

(D) Suppose we have two events, A and B. The spacetime coordinates of A and B observed by Ms. U and Mr. P are related by the Lorentz transformation. Subtracting one from the other, we obtained how the differences in space and time, $\Delta x = x_A - x_B$ and $\Delta t = t_A - t_B$, transform from one observer to the other, namely

$$\Delta t' = \frac{\Delta t + u \Delta x}{\sqrt{1 - u^2}} \quad \text{(temporal)}, \qquad \Delta x' = \frac{\Delta x + u \Delta t}{\sqrt{1 - u^2}} \quad \text{(spatial)}$$

(This is of course just a trivial rewrite of what we wrote before. Previously, we implicitly took event B to be when the origin of the two frames of reference coincided.)

(E) The Lorentz transformation actually specifies the geometry of spacetime.

Geometry is determined by the distance between two neighboring points. Consider the 2-dimensional plane studied by Euclid and by schoolchildren ever since, on which the two points have coordinates (x, y) and $(x + dx, y + dy)$ respectively, with dx and dy treated as infinitesimally small quantities, thus defining what we mean by neighboring. (In other words, dx is just the infinitesimal version of Δx.) Euclidean geometry is then specified by the Pythagoras theorem, which states that the distance dl between these two points in the plane, whose coordinates differ by (dx, dy), is given by $dl^2 = dx^2 + dy^2$, which is immediately generalized to $dl^2 = dx^2 + dy^2 + dz^2$ in 3-dimensional Euclidean space, and thence to D-dimensional Euclidean space for D any positive integer.

The distance between two points far apart could then be obtained by adding these infinitesimal distances along a straight line path between them. What is the point of doing this? For a Euclidean space, this is completely unnecessary. But when we moved to curved spaces, this is how geometry is defined, the point being that in an infinitesimally small neighborhood, reasonably smooth curved spaces look flat, as we will see later in this chapter.

Fine. Back to our discussion of spacetime. When we described Einstein's clock, we thought of the distance between the two mirrors and the light travel time between them as finite, but we could also think of these quantities as infinitesimally small. Thus, what we have discovered is that the distance, or separation, ds between two nearby spacetime points, with coordinates (t, x, y, z) and $(t + dt, x + dx, y + dy, z + dz)$, is given by the invariant

$$ds^2 = dt^2 - (dx^2 + dy^2 + dz^2)$$

(In other words, in the thought experiment with Einstein's clock, dt and dx correspond to T and X respectively, while $dy = 0$ and $dz = 0$. We found the invariant $T^2 - X^2$.) By now, you should realize also that if we do not use units with $c = 1$, we would have $ds^2 = (c\,dt)^2 - (dx^2 + dy^2 + dz^2)$.

In my textbook on Einstein gravity, I like to say that Whoever sets down the laws of the universe likes the Pythagoras theorem so much that He, She, They, or It wants to extend it to spacetime.

Thus, was it not for the crucial minus sign in ds^2, we would have ended up in a 4-dimensional Euclidean space rather than spacetime. The minus sign is the simplest[17] thing you can think of to distinguish space from time. Thus, spacetime is not Euclidean, but still flat, as was first pointed by Hermann Minkowski, who taught Einstein what would be regarded as lower level undergraduate math nowadays and whom Einstein referred to as an excellent teacher. By the way, Einstein said that this interpretation of "special relativity" as the geometry of a 4-dimensional spacetime had never occurred to him.

The opening words[18] of Minkowski's announcement in 1908 still resound across physics today: "Gentlemen! ... From now on, space by itself and time by itself must sink into the shadows, while only a union of the two preserves independence."

After gravity wave was detected, some journalists expressed surprise that it also travels at the speed of light.[19] In fact, the geometry of spacetime compels[20] any massless particle to travel with the speed c. In short, the speed of light c is not exclusively a property of light, but characterizes the geometry of spacetime.

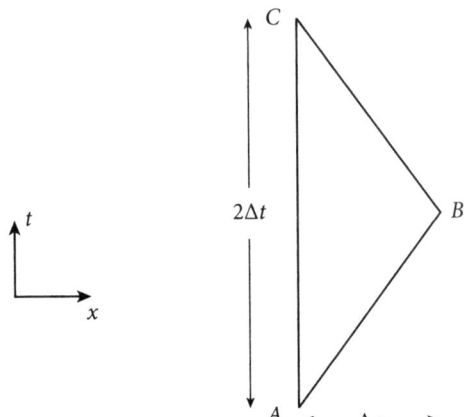

Figure 3. A triangle to illustrate the stark contrast between Euclidean space and Minkowski spacetime

(F) Indeed, the "Pythagorean character" of Einstein's spacetime could also be seen from the Lorentz transformation (in units with $c = 1$ for clarity), in the limit u much less than 1. Simply drop the square root factor from the Lorentz transformation to write $t' \simeq t + ux$, $x' \simeq x + ut$. Compare this with what happens if we rotate the x- and y-axes into each other through a small angle θ: $x' \simeq x + \theta y$, $y' \simeq y - \theta x$. Except for a crucial minus sign, we see that the Lorentz transformation amounts to a small "rotation" in spacetime.

(G) If $dt^2 > dx^2 + dy^2 + dz^2$, the separation between the two points in spacetime is said to be timelike. But if $dt^2 < dx^2 + dy^2 + dz^2$, then ds^2 becomes negative according to (E) and the separation between the two points in spacetime is said to be spacelike. At the transition, when $dt^2 = dx^2 + dy^2 + dz^2$, the separation is said to be lightlike. We and other lumbering giants in the classical world could only move in the timelike direction, traversing more time than space (in units with $c = 1$).

(H) What a difference a minus sign could make! The stark contrast between Euclidean space and Minkowski spacetime is underlined, as we will see presently, by a theorem taught to schoolchildren: the length of any one side of a triangle is less than the sum of the lengths of the other two sides. In other words, the shortest distance between any two points is a straight line.

Consider the triangle shown in Minkowski spacetime (figure 3) with the side AC having length $2\Delta t$ as indicated, with us using the words "length" and "separation" interchangeably as is customary in physics. The length of the side AB would have length $\sqrt{(\Delta t)^2 - (\Delta x)^2}$, and the same for the side BC. For AB and BC both timelike, $2\sqrt{(\Delta t)^2 - (\Delta x)^2}$ is manifestly smaller than $2\Delta t$, in violation of the theorem just cited. Indeed, if the separation between B and A (also, between B and C) is lightlike (so that the angle BAC is "equal to 45 degrees"), then the sum of the lengths of the two sides would equal zero, definitely less than $2\Delta t$.

(I) The bottom line: The 3-vector $\vec{x}$ is partnered with t, or if you prefer, with ct, to form the 4-vector x^μ, with the index $\mu = 0,\ 1,\ 2,\ 3$ and[21] $x^0 = t$, $x^1 = x$, $x^2 = y$, $x^3 = z$, that is $x^\mu = (x^0,\ x^1,\ x^2,\ x^3) = (t,\ x,\ y,\ z)$.

But as soon as that happens, then every concept in physics possessing a direction would also have to be promoted from a 3-vector to a 4-vector[22] in spacetime. This self-evident requirement rests on the very definition of dimension. That space is 3-dimensional means that we cannot have a fundamental concept in physics defined in terms of a 2-dimensional vector. Similarly, that spacetime is 4-dimensional behooves us to talk about 4-vectors rather than 3-vectors.

This apparently innocuous remark leads to some of the most amazing implications for physics, as we will see shortly.

No paradox, only misunderstanding

The so-called theory of special relativity is often taught as a series of paradoxes, but many theoretical physicists, myself included, consider that the wrong pedagogical approach. Instead, we should think of Einstein and Minkowski as uncovering the geometry of spacetime. The so-called paradoxes stem from misunderstanding.

Take time dilation. Ms. U sitting still (and hence $\Delta x = 0$) felt that time Δt has elapsed. Then Mr. P feels, according to "temporal" after we plug in $\Delta x = 0$, that $\Delta t' = \dfrac{\Delta t}{\sqrt{1 - u^2}}$, has elapsed. Since $\sqrt{1 - u^2}$ is less than 1, $\Delta t'$ is larger than Δt, a phenomenon known as time dilation.

An apparent paradox arises when students ask, "Aren't the two observers on equal footing? Ms. U thinks that Mr. P is the one who is moving, but Mr. P could also think that Ms. U is the one who is moving. How could Mr. P feel more time passing than Ms. U?" Yes, indeed. If we plug $\Delta x' = 0$ into "spatial," we obtain $\Delta x = -u\Delta t$, which when plugged into "temporal" gives $\Delta t' = \dfrac{\Delta t - u^2 \Delta t}{\sqrt{1 - u^2}} = \sqrt{1 - u^2}\,\Delta t$. Now $\Delta t = \dfrac{\Delta t'}{\sqrt{1 - u^2}}$ is larger than $\Delta t'$, as expected. So Ms. U also feels more time passing than

Mr. P, but as you can see, the "paradox" arises[23] because people are not comparing the same things: in one case, $\Delta x = 0$, in the other, $\Delta x' = 0$.

Similarly for length contraction and the so-called pole in the barn paradox. And the twin paradox, et cetera.[24]

How could there be any internal logical paradox left in spacetime physics? The theory, if you still insist on calling it a theory, is more than a century old, and is now being checked a zillion times a day. It may well be extended and modified in the future, but it shouldn't contain any logical inconsistency.

The loss of simultaneity implies local, not global, conservation

One of the most stunning consequences of the peculiar geometry of Einstein's spacetime is the loss of the commonsense notion of simultaneity.[25] Suppose Ms. Unprime observes events A and B occurring simultaneously. In other words, $t_A = t_B$, that is $\Delta t = 0$. What does Mr. Prime see? Simply plug in $\Delta t = 0$ into "temporal" to obtain $\Delta t' = \frac{u \Delta x}{\sqrt{1-u^2}}$, which is plain as pie not zero. Mr. Prime could swear that the two events did not occur simultaneously.

The loss of of simultaneity has important implications for the conservation laws discussed in chapter 5. To be definite, let us discuss the conservation of charge. Before Einstein, we could postulate that the conservation is global. Supposes a charge disappears in Berlin but instantaneously reappears in Timbuktu. Then charge conservation still holds: the total number of charges in the universe is unchanged.

But in Einstein's spacetime, with the loss of of simultaneity, mere global conservation no longer suffices. Another observer moving by could see the charge disappear, and not reappear till some time later. Or a charge could appear first out of the blue, and disappear somewhere else later. Thus, the loss of of simultaneity implies that conservation laws must be local, not merely global.

This requirement that physics be local practically demands the existence of fields in order to propagate information from one place in spacetime to somewhere else.

Summary: Time marries space, and simultaneity falls. Physics has to be local, and thus needs fields.

Temporal order reversal led physicists to antimatter

Even more dramatic than the loss of of simultaneity is that the temporal order of events could be reversed! Suppose Ms. Unprime observes event A to occur before B, that is $\Delta t = t_A - t_B$ is negative. Again, look at the "temporal" equation: $\Delta t' = \frac{\Delta t + u\Delta x}{\sqrt{1-u^2}}$. Since u and Δx could have either sign, the product $u\Delta x$ could certainly be positive, and if sufficiently positive, could render $\Delta t' = t'_A - t'_B$ positive. Thus, Mr. Prime could with equal justice insist that he observes event A to occur after B!

Let us examine the condition under which "something" could travel between A and B. Just a couple of lines of high school algebra would show that this is not possible. For $\Delta t + u\Delta x$ to be positive, we need $u\Delta x > -\Delta t$, and after squaring both sides, $u^2(\Delta x)^2 > (\Delta t)^2$. Since u^2 is less than 1, we have $(\Delta x)^2 > u^2(\Delta x)^2 > (\Delta t)^2$.

In English, to go between A and B, that thing would have to traverse more space than the time elapsed. (Remember, we are working in units with $c = 1$.) So the condition is travel faster than the speed of light, which is strictly forbidden by Einstein himself. (Note, however, that the relative velocity u between Ms. Unprime and Mr. Prime is still less than 1, as always.)

Now consider the scattering of a photon off an electron in the quantum world. We will depict the moment to moment happening by a Feynman diagram (see figure 4), in a totally straightforward (and theoretically obvious) manner, as described in the figure caption. Having gone through chapter 4, you are practically an expert on Feynman diagrams.

Look at the Feynman diagram in figure 4(a). (Time is shown running upward, with space extending horizontally.) An electron emits a photon at the spacetime point E, propagates from E to the spacetime point A, where and when it absorbs a photon and then goes on its merry way. This depicts the mundane process of an electron emitting and absorbing a photon, something that happens all around us zillions upon zillions of times every second. In particular, it is happening right now in your eyeball as we speak. As you can see (no "pun" intended), in this quantum process, an electron and a photon come in, and an electron and a photon go out.

Here comes the antielectron, ready or not

But now, consider an observer zipping by, such as that for her, the temporal order of E and A is reversed, as indicated in figure 4(b). She sees A

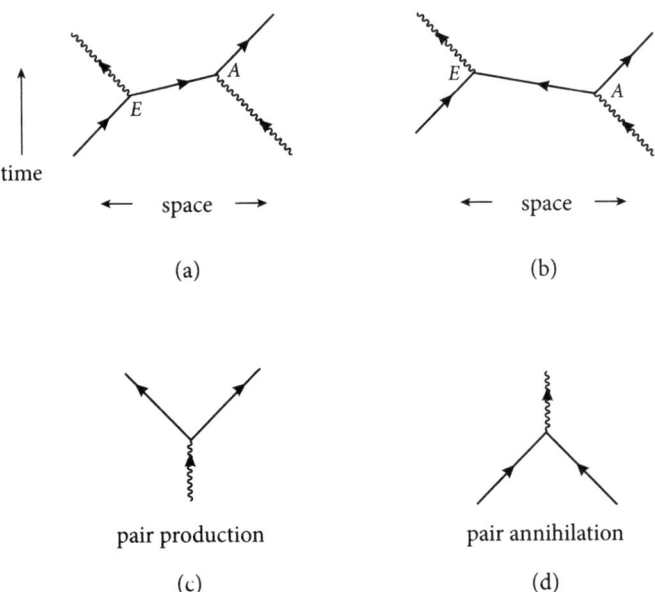

Figure 4. A quantum process described in the text in more detail. (a) An electron emits a photon at a point E in spacetime, then propagates from E to a point A later in time, where it absorbs a photon. (b) An observer zipping by sees E occurring later in time than A. She sees a photon producing an electron and a positron at A. The electron goes off somewhere, while the positron propagates from A to E, where it annihilates an incoming electron to form a photon. Two different observers are making different observations but subscribing to the same laws of physics. (c) What the observer in (b) sees at A is isolated for the reader's convenience: we have pair production, in which a photon produces an electron and a positron. (d) What the observer in (b) sees at E is isolated for the reader's convenience: we have pair annihilation, in which an electron and a positron annihilate each other and become a photon. This is the reverse of the process in (c).

occurring before E, not after! At the point A, a photon is absorbed and an electron goes away, but that's not all, something propagates from A to E. Let us look at this more carefully.

A photon, without any electric charge, comes in at the point A, at which an electron goes away, off to somewhere we don't know nor care about, but carrying a negative charge, and something else goes off. By charge conservation, this mysterious "thingy," hitherto unknown to physics, must carry away positive charge.

We have discovered the positron, a particle carrying a charge opposite to that carried by the electron, but otherwise the same as the electron. In particular, it has exactly the same mass[26] as the electron. The antiparticle of the electron crashes into physics! Here it comes, ready or not. By now, this astounding prediction, that every particle has an antiparticle with the same mass but opposite charges,[27] has been verified countless times to a high degree of accuracy.

I must caution you that what I just gave you is merely a heuristic argument. For a full fledged discussion, a proper formulation of relativistic quantum physics in terms of quantum field theory is necessary. The particle that propagates from E to A in figure 4(a) and from A to E in figure 4(b) is known as a virtual particle, in contradistinction to a physical particle. As we saw in chapter 4, in quantum field theory, particles are described as excitations in a field, and they propagate as waves, not as point particles as shown pictorially in Feynman diagrams. Indeed, Feynman diagrams, drawn with lines (both straight and wavy), could be occasionally misleading[28] and must be interpreted with care.

Let us continue looking at figure 4(b). The positron created at A continues on to E where it meets an incoming electron. Both the positron and the electron disappear, and a photon is emitted at E.

If we examine the vertex (as it is called in the jargon) A in isolation, we see (figure 4(c)) a photon coming in and producing an electron and a positron, known as pair production. The reversed process (figure 4(d)), in which an electron and a positron annihilate each other and become a photon, also occurs. Indeed, if you now look at the vertex E in isolation, you see pair annihilation happening right before your eyes. Pair production and pair annihilation of matter and antimatter may sound like science fiction, but in fact occur routinely in accelerators and in cosmic rays.

Again, for the benefit of the "truth is relative" crowd, let us repeat Einstein's point: While two observers could describe a photon scattering off an electron differently, they must subscribe to the same laws of physics. Reconciling their apparently different descriptions, physicists discovered antimatter.

If you examine this argument that the universe must contain antimatter as well as matter, you would see that it involves the interplay of both relativistic and quantum physics. The word "both" is crucial here. A nonrelativistic quantum physicist could shake and bake the Schrödinger equation till smoke comes out her ears and she would never see an antiparticle (as I had said before). Neither would a relativistic classical physicist ever encounter antimatter.

Finally, I want to emphasize a truly important point. Did you see anything going backward in time? I hope not. Nothing was going backward in time! We are just talking about two observers with different interpretations of the same process. If the reader would indulge me, I would like to quote from my textbook on quantum field theory, in a chapter about quantizing the Dirac field: "Did I speak of an electron going backwards in time? Did I mumble something about a sea of negative energy electrons? This metaphorical language, when used by brilliant minds, the likes of Dirac and Feynman, was evocative and inspirational, but unfortunately has confused generations of physics students and physicists."[29]

The most famous formula in physics: $E = mc^2$

Understandably, many among the intellectually curious would like to know how the most famous formula in physics $E = mc^2$ pops out. I offer those readers the simplest possible derivation of this iconic formula that I am able to devise.

To follow this derivation, you need to have a rudimentary knowledge of Newtonian mechanics, namely that a particle of mass m moving with velocity $\vec{v}$ has momentum equal to $\vec{p} = m\vec{v}$ and kinetic energy equal to $E = \frac{1}{2}mv^2$. You don't even have to know the $\frac{1}{2}$; I only need you to note that while momentum has dimension of mass times velocity $[p] = ML/T$, energy has dimension of mass times velocity squared. Thus $[E] = M(L/T)^2 = [p](L/T)$, and energy has the dimension of momentum times a velocity. In other words, momentum has the dimension of energy divided by a velocity.

In Newtonian mechanics, a particle at rest (that is, with $\vec{v} = 0$) has no momentum and no energy.

"No!" said Einstein. A particle just sitting there has energy $E = mc^2$, which is enormous compared to the Newtonian energy of $E = \frac{1}{2}mv^2$ for any conceivable everyday value of v.

Most remarkably, to derive this famous formula, we barely have to know anything about the Lorentz transformation! All we have to know is that the Lorentz transformation turns x and ct into each other. In fact, all we need is how x transforms for u much less than the speed of light, namely for u in the Newtonian regime: $x' = x + ut = x + \frac{u}{c}ct$. But this is merely the everyday commonsense definition of the velocity u that Galileo wrote down almost four hundred years ago.

All this is so Newtonian that you may be wondering, where is c? Ah, in writing the second form for x' we are insisting the ct is what x transforms

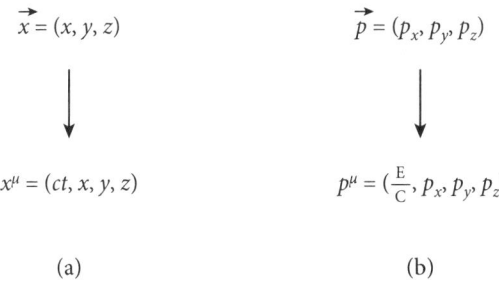

$$\vec{x} = (x, y, z) \qquad\qquad \vec{p} = (p_x, p_y, p_z)$$

$$x^{\mu} = (ct, x, y, z) \qquad\qquad p^{\mu} = \left(\tfrac{E}{c}, p_x, p_y, p_z\right)$$

(a)   (b)

Figure 5. Searching for a partner! To qualify to live in spacetime, a spatial vector, known as a 3-vector, must find some quantity to partner with in order to form a spacetime vector, known as a 4-vector. (a) The three familiar spatial coordinates partner with time. (b) The three components of momentum (that is, the component of the momentum in the x, y, and z directions) partner with energy. Dimensional considerations dictate the factors of c. Thus, ct has the same dimension as length, and E/c the same dimension as momentum.

into, not t. So, that's how c sneaks into this derivation of the famous formula.

Now that we understand Einstein's so-called special relativity is about the geometry of 4-dimensional spacetime and know that the 3-vector $\vec{x}$ is partnered with ct to form the 4-vector x^{μ}, we realize that every 3-vector in Newtonian physics must partner with someone to form a 4-vector, as I said earlier. In 4-dimensional spacetime, 3-vectors are not allowed: they make no sense.

So, let's picture all these 3-dimensional concepts out wandering around looking for partners. See figure 5.

The natural partner for the momentum 3-vector $\vec{p}$ is energy. But it can't be just energy: I just showed you that momentum has the dimension of energy divided by a velocity. Partners must have the same dimension: you can hardly partner a kilogram with a meter, for example. The only velocity available to divide the energy by is c. So, just like the 3-vector $\vec{x}$ partnering with ct to form the 4-vector x^{μ}, the 3-vector $\vec{p}$ partners with E/c to form a 4-vector p^{μ}.

To repeat, to dance together in spacetime, the two partners much have the same dimension. A meter cannot dance with a second! Neither can a momentum dance with an energy!

Back to Ms. Unprime on the train and Mr. Prime on the ground. How does $\vec{p}$ transform? Easy, the same as $\vec{x}$. (The y and z components of $\vec{p}$ do not transform, just like the coordinates y and z, so we can forget about them.)

The x component of $\vec{p}$, call it simply p, transforms just like the coordinate x, namely $x' = x + \frac{u}{c}ct$. To figure out what p' is, we simply copied what we just wrote for x', and then substitute accordingly. An explicit substitution list might help some readers: $x \to p$, $ct \to \frac{E}{c}$, and $x' \to p'$. (Here, arrows mean substitution.) Thus

$$x' = x + \frac{u}{c}ct \quad \to \quad p' = p + \frac{u}{c}\left(\frac{E}{c}\right) = p + \frac{uE}{c^2}$$

Almost there! Ms. U, instead of tossing a ball forward as when we derived the addition of velocities, holds the ball in her lap. Since Mr. P sees the ball moving forward with velocity u, he thus concludes that it has momentum $p' = mu$. (Notice that this statement is purely Newtonian since we are assuring that u is much less than c. Everything, the train, the ball, and Ms. U are all moving slowly relative to Mr. P.) But to Ms. U, the ball is at rest, and so has momentum $p = 0$ and energy $E = 0$. So the equation we just wrote down, $p' = p + \frac{uE}{c^2}$, says that $mu = 0 + 0$, which is totally wrong.

How can we fix this? Evidently, p and E cannot both be zero. Well, $p = 0$: the ball is not moving and so obviously has momentum equal to 0. The only way out is for E not to be zero. Furthermore, we see that the left hand side of this equation is proportional to u, and we see a u on the right hand side multiplying E. Thus, instead of the inconsistent $mu = 0 + 0$, we are forced to $mu = 0 + \frac{uE}{c^2}$.

Canceling off the u and multiplying by c^2, we obtain, ta dah,

$$E = mc^2$$

Trumpets please! An object of mass m just sitting there at rest has energy, and an enormous amount at that! As you and I know, that was a world changing observation Einstein made.[30]

That is your promised reward. You can now show others how $E = mc^2$ emerges.

Interestingly, although Lorentz found his eponymous transformation, he was not able to work out its ramifications for physics. It is tempting to hypothesize that this is because Einstein was young at the time while Lorentz was old.

A summary of how Einstein obtained the most celebrated formula of physics:

How x transforms according to Lorentz: $x' = x + \frac{u}{c}(ct)$.
How momentum transforms according to Lorentz: $p' = p + \frac{u}{c}\left(\frac{E}{c}\right)$.
The ball at rest in Ms. U's lap has $p = 0$, while Mr. P sees the ball moving by with $p' = mu$.
Comparing with how momentum transforms, we obtain $mu = 0 + \frac{u}{c}\left(\frac{E}{c}\right)$.
Thus, E cannot be zero as in Newtonian physics. Einstein was forced to $E = mc^2$.

(To repeat, the second line follows from the first line by using the substitution rules given just now.)

The Lord did not lead him around by the nose

Incidentally, Einstein didn't have "the most famous formula in physics" in his 1905 paper proposing a new architecture of spacetime. This famous relation appeared only a few months later in a brief note.[31]

At this point, I can't resist showing you the relevant page from Einstein's paper. See figure 6. A puzzle for you: Can you find $E = mc^2$? Okay, a hint. Einstein did not denote the speed of light by c, but by V.

See it now? Einstein considered an object moving slowly with velocity v emitting an electromagnetic wave carrying away an amount of energy equal to L. The kinetic energy of the object before and after emission are denoted by K_0 and K_1 respectively. He obtained the relation $K_0 - K_1 = \frac{L}{V^2}\frac{v^2}{2}$.

Oy vey! Let's translate this into modern notation. Denote the mass of the object before and after emission by m_0 and m_1 respectively, so that $K_0 = \frac{1}{2}m_0 v^2$ and $K_1 = \frac{1}{2}m_1 v^2$ respectively. Then the central equation in Einstein's paper would read in more modern notation

$$(m_0 - m_1)\frac{v^2}{2} = \frac{E}{c^2}\frac{v^2}{2}$$

Canceling and cleaning up, we see that $E = \Delta mc^2$, with $\Delta m = m_0 - m_1$ the change in mass due to the emission of energy E. That's it!

that is independent of the properties of the body. Furthermore, the difference $K_0 - K_1$ depends on the velocity in the same way as does the kinetic energy of an electron (*loc. cit.*, sec. 10).

Neglecting magnitudes of the fourth and higher order, we can get[1]

$$K_0 - K_1 = \frac{L}{V^2} \frac{v^2}{2}.$$

From this equation one immediately concludes:

If a body emits the energy L in the form of radiation, its mass decreases by L/V^2. Here it is obviously inessential that the energy taken from the body turns into radiant energy, so we are led to the more general conclusion:

The mass of a body is a measure of its energy content; if the energy changes by L, the mass changes in the same sense by $L/9 \cdot 10^{20}$ if the energy is measured in ergs and the mass in grams.

It is not excluded that it will prove possible to test this theory using bodies whose energy content is variable to a high degree (e.g., radium salts).

If the theory agrees with the facts, then radiation carries inertia between emitting and absorbing bodies.

(*Annalen der Physik* 18 [1905]: 639–641)

Figure 6. The most famous formula of physics written in the "antiquated" notation of 1905. So now you know, all those cartoons showing for example Einstein writing $E = ma^2$, $E = mb^2$, $E = mc^2$, etc., on the blackboard and pondering which ones to cross out are not quite accurate. Page reproduced from "Does the Inertia of a Body Depend on Its Energy Content," by Albert Einstein (*Annalen der Physik* 18 [1905]: 639–641). Translated in *The Principle of Relativity* by W. Perrett and G. B. Jeffrey (Dover, 1923).

I like how Einstein wrote, with due modesty, "It is not excluded that . . ." Such a far cry from the modern "My theory can explain everything!"

Indeed, Einstein was apparently a bit worried about being seduced by his own argument, and wrote to a friend: "One more consequence of the paper on electrodynamics has also occurred to me. . . . The argument is amusing and seductive; but for all I know the Lord might be laughing over it and leading me around by the nose."

As we all know, the Lord did not lead him around by the nose.

The mass of an elementary particle measures its inability to move at the speed of light

By now you could see why physicists who work in this area habitually set $c = 1$. Keeping c around merely adds to the clutter. If we did that, we would have obtained that for an object of mass m at rest, its energy E equals m. We could then put c back in, if so desired, by following the procedure outlined earlier.

Here is an even quicker derivation. Given that $t^2 - \vec{x}^2$ is an invariant and that $(E, \vec{p})$ forms a 4-vector just like $(t, \vec{x})$, then $E^2 - \vec{p}^2$ is also an invariant. As an invariant, this quantity is an intrinsic characteristic of the particle, which we are going to call the square of the mass m of the particle. In other words, $E^2 - \vec{p}^2 = m^2$, or more commonly written in particle physics as

$$E^2 = \vec{p}^2 + m^2$$

The famous formula, $E = mc^2$, or $E = m$ in units with $c = 1$, is "merely" a special case of this relation between the energy, the momentum, and the mass of a particle, when its momentum $\vec{p}$ equals zero, that is, when it is at rest.

It is an understatement that the formula $E^2 = \vec{p}^2 + m^2$ has been verified zillions of times, nay, it has been verified a gazillion times since Einstein wrote it down.

Nowadays, when a hitherto unknown particle is discovered, experimentalists are hardly going to grab it and put it on a scale to weigh it and deduce its mass. Rather, they measure its energy E and its momentum $\vec{p}$ and declare its mass to be the m given by the relation $m^2 = E^2 - \vec{p}^2$. This in fact provides a test of Einstein's theory; every time experimentalists see this particle, its energy and its momentum may be different, but its mass better be the same, that is, an invariant, in constant rebuttal of the disciples of "relative truth."

Look at this another way: $\vec{p}^2 = E^2 - m^2$. You could think of the magnitude p of the momentum desperately trying to be equal to the energy E, but always falling short by an amount measured by m. As you could see, p starts to approach E only when they both become enormous compared to m, in which case m becomes negligible. In Einstein's theory, the velocity of a particle is defined to be the ratio p/E. Thus, since $p = E$ for massless particles, they all move with speed equal to 1, that is, equal to c. Massive particles approach that speed at high energies, but could never exceed it.

You could say that mass of a particle measures its inability to move at the speed of light. By this criterion, we are lumbering masses in spacetime.

An executive interim summary

Time for an executive summary!

The Galilean addition of velocities is just everyday common sense.
The speed of light stems from some private business between the electric and magnetic fields and hence contradicts the Galilean transformation.
The Lorentz transformation generalizes the Galilean transformation but yet must reduce to it in everyday situations.
The Lorentz transformation forces space and time to transform into each other, and hence momentum and energy to transform into each other.
It immediately follows that $E = mc^2$, even without invoking any details about the Lorentz transformation.

From the exterminator of relativity to the choreographer of spacetime

After discovering the geometry of spacetime, Einstein spent the next ten years searching for, and finding, a theory of gravity, now more commonly known as general relativity, in an unfortunate extension of the unfortunate term special relativity.

Why does Einstein have to go off on this arduous quest? The point is that gravity lives in spacetime and hence must conform to the architecture of spacetime.

Again many roads lead to Einstein gravity, and I will mention only a couple here.

Einstein's principle of equivalence

From contemplating inertial frames moving at a constant velocity relative to each other, Einstein naturally graduated to considering noninertial frames. From everyday experience, we know that if the car we are sitting in suddenly brakes, we will be thrown forward. And if the car suddenly accelerates, we will feel a force pushing us into our seats. If the speeding car rounds a corner, we will feel a centrifugal force throwing us outward.

Einstein was thus inspired to formulate his equivalence principle, which states that in a sufficiently small region of spacetime, an accelerating frame and a gravitational field are indistinguishable from each other, as far as physical effects are concerned.[32]

It is important to realize that this is quite different from our earlier discussion, which insists that the two observers see the same laws of physics. Here the observer in an inertial frame, such as a frame at rest, watching the accelerating frame goes by, does not feel a gravitational field, while the observer in the accelerating frame would insist that a gravitational field is present.

General coordinate transformation

When we transform from one inertial frame to another, the geometry of Minkowski's spacetime, defined by $ds^2 = dt^2 - (dx^2 + dy^2 + dz^2)$, remains unchanged. Indeed, the Lorentz transformation is carefully constructed in order to leave ds^2 invariant. But when we transform to an accelerating frame, Minkowski's ds^2 is sure as hot cash going to turn into something much more involved.

In fact, ds^2 will involve not only dt^2, dx^2, dy^2, and dz^2, but also cross terms such as $dtdx$, $dtdy$, $\cdots$, $dydz$. Furthermore, each of these terms will have coefficients that are functions of t, x, y, z (instead of the babyish $+1$, -1, and 0).

Let me give you a handwaving account of how that would come about. Go back to Ms. Unprime and Mr. Prime. If the train accelerates, then $x' = x + ut + \frac{1}{2}at^2$, in the simplest Newtonian example of constant acceleration a. In other words, x' is no longer a simple linear function of t. Then $dx'^2 = (dx + udt + atdt)^2$. Upon squaring we see two new features: the coefficient of dt^2 now equals $(u + at)^2$ and is no longer just 1, and a cross term $dxdt$, with coefficient also dependent on t, appears.

Indeed, more generally, x' may equal some arbitrary function of not only x and t, but also of y and z. Similarly, t', y', and z' all may equal arbitrary functions of t, x, y, z. (This is known as a general coordinate transformation.) Then dt', dx', dy', dz' and dt, dx, dy, dz are no longer related simply by the Lorentz transformation.

Hence Minkowski spacetime surely cannot hold in general. Let's say it holds in Mr. Prime's frame, so that $ds^2 = dt'^2 - (dx'^2 + dy'^2 + dz'^2)$, but once we transform to some other frame, it would no longer be a simple "sum" of four terms with one coefficient equal to 1 and three equal to -1.

The coefficients would all become arbitrary functions of t, x, y, z, with cross terms popping up all over the place.

Clearly, it would be advantageous to use, instead of t, x, y, z, the index notation we talked about earlier, and write x^μ with $\mu = 0$, 1, 2, 3. Then $ds^2 = g_{00}(x)(dx^0)^2 + g_{11}(x)(dx^1)^2 + \cdots + 2g_{01}(x)dx^0 dx^1 + 2g_{02}(x)dx^0 dx^2 + \cdots + 2g_{23}(x)dx^2 dx^3$. There are ten terms on the right hand side (to test if you are following, count them!) altogether but I have not bothered to write them all out; the ones I did not write out are indicated by dots. Einstein then invented what is known in physics as the Einstein repeated index summation and simply wrote

$$ds^2 = g_{\mu\nu}(x)dx^\mu dx^\nu$$

with the stipulation that indices that are repeated, namely μ and ν, are to be summed over, ranging through 0, 1, 2, 3. The ten functions $g_{\mu\nu}(x)$ (namely $g_{00}(x)$, $g_{01}(x)$, $\cdots$, $g_{33}(x)$) are known collectively as the spacetime metric: as the jargon suggests, they measure spacetime.

A potential confusion about the notation used by Einstein. The notation $g_{\mu\nu}(x)$ is shorthand for $g_{\mu\nu}(x^0, x^1, x^2, x^3)$. The letter x is used to denote x^0, x^1, x^2, x^3 collectively. (There are only twenty-six letters in the alphabet!) In other words, each of the ten functions $g_{\mu\nu}(x)$ is a function of t, x, y, z, that is, a function of spacetime. (Indeed, it would be absurd to say that they are functions of the coordinate x only. What is so special about the x direction?)

To make sure that we understand Einstein's compact notation, let us see how flat Minkowski spacetime appears as a special case, with a particularly simple form of $g_{\mu\nu}(x)$. The ten functions actually are not functions, just numbers, and all but four are equal to 0. These four are $g_{00} = +1$, $g_{11} = -1$, $g_{22} = -1$, $g_{33} = -1$. In other words, $ds^2 = (dx^0)^2 - (dx^1)^2 - (dx^2)^2 - (dx^3)^2$, which I trust you to recognize as flat Minkowski spacetime written using indices rather than using t, x, y, z. Since the earth's gravitational field is so weak, most of the time we are hanging out in a spacetime that is very very close to flat Minkowski spacetime. Thus, flat Minkowski spacetime is by far the most important spacetime to know and love. Not surprisingly then, theoretical physicists traditionally assign the Minkowski metric, that is, the metric I just described, a special symbol, namely $\eta_{\mu\nu}$, using the Greek letter η (pronounced eta). Nothing profound here: we merely define $\eta_{\mu\nu}$ by specifying that the only non-zero components of η are $\eta_{00} = +1, \eta_{11} = -1, \eta_{22} = -1$,

$\eta_{33} = -1$. In other words, we can write the flat spacetime more compactly as $ds^2 = \eta_{\mu\nu} dx^\mu dx^\nu$.

From curved space to curved spacetime

It is essential to realize that in Einstein's foundational insight about gravity, the two observers, Ms. Unprime and Mr. Prime, do not observe the same physics, in contrast to our earlier discussion about the so-called special relativity. One feels gravity, the other not.

In summary, the heuristic motivation given in the preceding paragraph leads us to consider general[33] spacetimes described by ten arbitrary functions $g_{\mu\nu}(x)$ of the location in spacetime specified by the coordinates x. The corresponding spacetimes are in general curved!

That spacetime is curved is perhaps not so surprising if I remind you of the curved space we are most familiar with, namely the surface of the globe we live on. Physicists and mathematicians traditionally denote latitude[34] and longitude by the Greek letters θ and φ respectively. First, consider two nearby points with the same longitude φ but slightly different latitude, θ and $\theta + d\theta$. Then the distance between them equals $d\theta$. (We are using the radius of the globe as our unit of distance.) Next, consider two nearby points with the same latitude θ but slightly different longitude, φ and $\varphi + d\varphi$. The distance between them is now given by $f(\theta)d\varphi$ and depends on the latitude θ, as you can see from figure 7. The function $f(\theta)$ is equal to 1 at the equator, but decreases steadily as we move away from the equator and vanishes at the two poles. (In fact, $f(\theta) = \sin\theta$, take my word for it.)

Thus, on a sphere, the distance ds between two neighboring points, one with coordinates (θ, φ) and the other with coordinates $(\theta + d\theta, \varphi + d\varphi)$, is given by $ds^2 = d\theta^2 + \sin^2\theta \, d\varphi^2$. This is a 2-dimensional space with coordinates $x^1 = \theta$ and $x^2 = \varphi$, and metric $g_{11} = 1$, $g_{12} = g_{21} = 0$, $g_{22} = \sin^2\theta$. That the metric varies with location indicates that the space is curved.

As another example of Nature's kindness to physicists, two of the greatest German mathematicians of the 19th century, Carl Friedrich Gauss and Bernhard Riemann, had already worked out the mathematics of curved spaces, a subject known as differential geometry. In particular, Riemann produced an expression now called the Riemann curvature tensor, which involves differentiating the metric twice. All you have to do, when given a metric,[35] is to plug it into Riemann's expression, and you know what the curvature of the space described by that metric is. Riemann did all the work for us!

North pole

Figure 7. The distance ds between two neighboring points on the sphere, one with coordinates (θ, φ) and the other with coordinates $(\theta + d\theta, \varphi + d\varphi)$, is given by $ds^2 = d\theta^2 + (f(\theta))^2 \, d\varphi^2$, as explained in the text. The function $f(\theta)$ is equal to 1 at the equator, but decreases steadily as we move away from the equator and vanishes at the two poles. Reproduced with permission from *On Gravity* (Princeton University Press, 2018).

Furthermore, the work of Gauss and Riemann on curved spaces[36] could be lifted almost instantly to curved spacetime. Nature's kindness notwithstanding, Einstein was unfortunately not acquainted with differential geometry, and his lack of the relevant mathematical knowledge[37] is one of several reasons that he had to struggle for ten arduous years to arrive at his theory of gravity.

To go from curved space to curved spacetime is rather simple. Just as in going from flat Euclidean space to flat Minkowski spacetime, we simply have to distinguish the time coordinate from the space coordinates by the judicious insertion of some minus signs.

To give you a flavor of what is involved, let me show you three famous spacetimes.

An expanding universe

Consider the spacetime described by

$$ds^2 = dt^2 - a^2(t)(dx^2 + dy^2 + dz^2)$$

Note that the 3-dimensional space defined by $(dx^2 + dy^2 + dz^2)$ is still Euclidean flat space, but because of the factor $a^2(t)$, the distance between two points is changing, either expanding or contracting. The math is almost immediate and elementary: consider two neighboring points in spacetime, (t, x, y, z) and $(t, x + dx, y, z)$, then the distance squared between them is given by $a^2(t)dx^2$ and so the distance between them equals $a(t)dx$ and changes with time. In this simple spacetime, which incidentally describes the universe we live in quite accurately, the distance between two possibly far separated points, with coordinates (t, x, y, z) and $(t, x + \Delta x, y, z)$, is given just by adding up the infinitesimal distances $a(t)dx$ to give $a(t)\Delta x$. Thus, if the function $a(t)$ increases with time, the universe is expanding.

People often ask, "It's that simple?" Yes, it's that simple.

Interestingly, while space is flat, spacetime is curved. Compare and contrast with flat Minkowski spacetime.

The hard part is to find the equation that determines the function $a(t)$, known as the scale factor of the universe. Well, Einstein found that equation and now just about any reasonably bright undergraduates can solve[38] for $a(t)$. (Incidentally, the so-called Hubble "constant" is just the fractional rate at which $a(t)$ is changing.[39]) Observational cosmologists then have the task of measuring $a(t)$. The behavior of $a(t)$, which determines whether the expansion is accelerating or decelerating, about which you have surely heard, depends on the ingredients filling the universe, in particular the relative proportion of dark energy and dark matter.[40] See figure 8.

That the metric describing an expanding universe is so simple is of course due to our restricting ourselves to cosmological scales, in which local blemishes such as a few galaxies here and there have been smoothed over. But that is how cosmology is done, at least in a first pass.

A black hole

Next, I will show you, ta dah, the spacetime that describes a black hole. Here it is: the curved spacetime around, and outside, a spherical mass distribution with mass M and radius R is described by[41]

$$ds^2 = \left(1 - \frac{2GM}{r}\right) dt^2 - \frac{1}{\left(1 - \frac{2GM}{r}\right)} dr^2 - r^2(d\theta^2 + \sin^2\theta \, d\varphi^2)$$

Here G denotes Newton's gravitational constant. This solution of Einstein's field equation was obtained in 1915, the very same year that Einstein

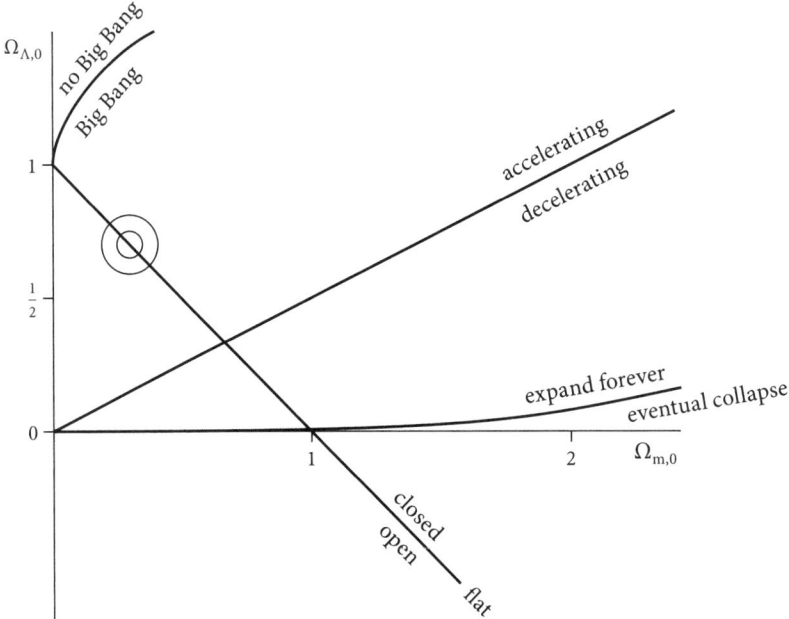

Figure 8. The cosmic struggle between dark energy and dark matter: a diagram describing the overall history of the universe according to how much dark matter and how much dark energy the universe contains at present. The amount of dark matter (denoted by $\Omega_{m,0}$ by the pros) is plotted along the horizontal axis, the amount of dark energy (denoted by $\Omega_{\Lambda,0}$) along the vertical axis. For example, a universe with $\Omega_{m,0} = 2$ and $\Omega_{\Lambda,0} = \frac{1}{2}$ will expand forever but with a decelerating rate. Our universe is observed to be within the circles with various degrees of confidence level. Reproduced with permission from *GNut* (Princeton University Press, 2013).

published his theory, by Karl Schwarzschild, an officer serving in the German army on the Russian front during World War I.[42] Interestingly, Einstein only found an approximate solution valid for large r, which was in fact adequate for his purpose of working out the three celebrated observational tests of his theory. He was elated when he learned, from a letter Schwarzschild wrote to him, that his highly nonlinear field equations had such a simple solution.

Some lay readers might be "blown away" by all these symbols, but in fact, ever since Einstein, physicists and physics students have had the opposite reaction: they were "blown away" by how amazingly simple this metric is!

Note that of the ten functions $g_{\mu\nu}$ of spacetime, no fewer than six vanish. Furthermore, as you could see from our discussion just now, the angular part of the spacetime $d\theta^2 + \sin^2\theta d\varphi^2$ just describes a sphere. Thus, the only nontrivial pieces of this metric are g_{tt} and g_{rr}, and not only that, instead of depending on all four of the coordinates t, r, θ, φ, these two guys depend only on r. And not only that, only the very simple function $(1 - \frac{2GM}{r})$ is involved. Isn't Nature kind?

Notice that at the beginning of this section I was careful to say "the curved spacetime around, and outside, a spherical mass distribution with mass M and radius R." Consider a star. The first thing you might have noticed is that R doesn't even come into the metric. The reason is clear: the solution is strictly for outside the star, that is, for $r > R$. That's how R sneaks in. For $r < R$, inside the star, hot gas whirls around, and the metric would depend on the detailed composition of the star and the conditions in its interior such as temperature and pressure. The metric just given is simply not valid for $r < R$.

The spatial part of the Schwarzschild metric is for sure not flat, in contrast to the expanding universe we just looked at. All the crazy science fiction sounding scenarios about travelers falling into a black hole and never being able to come out again and spacetime being warped stem from the factor $(1 - \frac{2GM}{r})$, appearing once as itself and once as its reciprocal $1/(1 - \frac{2GM}{r})$ in front of dt^2 and dr^2 respectively.

The key point is that as r, the radial distance from the origin $r = 0$, decreases from being larger than $2GM$ to being smaller, the ratio $\frac{2GM}{r}$ changes from smaller than 1 to larger than 1, and so these two factors flip sign, going from plus to minus: $(1 - \frac{2GM}{r}) > 0$ for $r > 2GM$, and < 0 for $r < 2GM$. Effectively, the time coordinate t and the radial coordinate r switch roles! In other words, r becomes the new time variable, while t becomes the new radial variable. All the weird stuff about black holes comes from this simple sign flip.[43]

The weird goings-on happen at $r = 2GM \equiv r_S$, called the Schwarzschild radius, or also, the horizon. So, we conclude that if $r_S < R$, that is if the horizon is inside the star, then nothing weird could happen: the Schwarzschild metric is simply not valid there. We have a star. But if $r_S > R$, then we have a black hole! The temporal coordinate t and the spatial coordinate r exchange roles at the horizon outside the black hole. Spacetime is no longer what we are used to!

By the way, you may have heard the common misconception that there is some "mysterious force sucking stuff into a black hole." No such force besides our old friend gravity!

How to recover Newtonian gravity?

I already emphasized earlier in this chapter that Newtonian physics is not overthrown by Einsteinian physics. Indeed, as I said earlier, in physics one stringent requirement of a successful new theory is that it must go over into the old theory under the appropriate circumstances. Some readers may know that in Newtonian gravity the energy of attraction between two masses of mass M and m separated by a distance r is given by $E = -GMm/r$. I like to show them how this pops out of the Schwarzschild metric. The "trick" is to put the speed of light c back by doing some simple dimensional analysis. Since we know that mc^2 is also an energy, we have $[GMm/r] = [mc^2]$. Canceling off m, we conclude that $[GM/rc^2] = 1$, in other words, the combination GM/rc^2 is dimensionless and could be subtracted from 1. Thus, in units with c not equal to 1, we have

$$ds^2 = \left(1 - \frac{2GM}{rc^2}\right)c^2 dt^2 - \frac{1}{\left(1 - \frac{2GM}{rc^2}\right)}dr^2 - r^2(d\theta^2 + \sin^2\theta\, d\varphi^2)$$

Note that we have also replaced t by ct (year by light year so to speak).

Now let c go to infinity so that the fraction $\frac{2GM}{rc^2}$ tends to zero. Behold, the spacetime becomes

$$ds^2 = c^2 dt^2 - (dr^2 + r^2 d\theta^2 + r^2\sin^2\theta\, d\varphi^2) - \frac{2GM}{r}dt^2$$

$$= c^2 dt^2 - (dx^2 + dy^2 + dz^2) - \frac{2GM}{r}dt^2$$

$$= ds^2_{\text{flat}} - \frac{2GM}{r}dt^2$$

Some readers might recognize that $dr^2 + r^2 d\theta^2 + r^2\sin^2\theta\, d\varphi^2$ is just $dx^2 + dy^2 + dz^2$ written in spherical coordinates.

So, you see that in the limit $c \to \infty$, the Schwarzschild spacetime ds^2 reduces to good old flat Minkowski spacetime ds^2_{flat} modified by adding the tiny amount $(-\frac{2GM}{r}dt^2)$.

Gravitational redshift

That tiny amount accounts for Newtonian gravity. We see with our own eyes Newtonian gravity emerging. In other words, a spacetime endowed with a Newtonian gravitational potential ϕ is described by

$$ds^2 = \eta_{\mu\nu}dx^\mu dx^\nu + 2\phi dt^2 \quad \text{Newtonian gravity with } \phi \simeq -\frac{GM}{r}$$

Viewed this way, we see another dramatic consequence of Einstein's theory: gravitational redshift. The presence of a gravitational field affects the rate at which clocks run. This causes the light emitted by massive stars to shift toward the red end of the electromagnetic spectrum. Indeed, the effect could be calculated almost instantaneously: dt^2 in the familiar flat spacetime is corrected to $(1 + 2\phi)dt^2$. In other words, dt is corrected to $\sqrt{(1+2\phi)}dt \simeq (1+\phi)dt$, with the approximation due to ϕ being much smaller than 1 in Newtonian gravity.

Of the three so-called classic tests of Einstein's theory, that gravity warps time was the last to be verified, in a terrestrial experiment performed at Harvard in 1960 by R. V. Pound and collaborator. Pound would joke that that he learned the true meaning of gravity while lugging the heavy experimental components up and down the tower of the Harvard physics department.

Our discussion here also leads to the somewhat cryptic quip[44] that whereas Newton curved time, Einstein also curved space.

Gravity waves

Humans first detected a gravity wave in late 2015. Now you know how to write down a gravity wave[45] propagating in Minkowski spacetime, namely $g_{\mu\nu}(x) = \eta_{\mu\nu} + h_{\mu\nu}(x)$ with $h_{\mu\nu}(x)$ treated as a small perturbation to $\eta_{\mu\nu}$, that is, as a ripple. When we insert this metric into Einstein's field equation, which we will discuss presently, we find the equation governing the wave amplitude $h_{\mu\nu}(x)$, an equation which is only slightly more involved than the equation governing an electromagnetic wave, which all physics undergraduates study. In particular, calculating the power in a gravity wave emitted by two black holes orbiting around each other[46] is not much more difficult than calculating the power in the electromagnetic wave emitted by two electric charges orbiting around each other. A reasonably bright undergraduate could calculate[47] the gravity wave emitted by two black holes orbiting around each other.

Einstein's field equation

> Compared with understanding gravity, the special theory of relativity
> was mere child's play.
>   —Einstein writing to Arnold Sommerfeld in 1912

I just gave you three examples of some interesting spacetimes, each defined by a set of $g_{\mu\nu}(x)$, but I have yet to tell you about Einstein's equation for determining the metric $g_{\mu\nu}(x)$. And that's what Einstein's decade-long struggle was about. If entire books can be written about the so-called special relativity, then entire tomes can be written about general relativity, which is what Einstein gravity is generally known as, as I have said. I have no choice other than give you a flavor of how one particular approach to Einstein gravity would go.[48]

Start with Newtonian gravity, which has this schematic form

Newtonian gravity
spatial variation of the gravitational potential or field ϕ = mass per unit volume

A distribution of mass in space generates a gravitational field ϕ determined by this equation (as I mentioned in chapter 4).

The approach I am going to describe is to generalize this equation to make it compatible with various symmetry requirements and with the equivalence principle. This equation is, of course, invariant under rotations; Newton himself noted that gravity does not pick out a special direction. But it is clearly not invariant under Lorentz transformations. Look at the left hand side of Newton's equation. Now that space and time are unified, spatial variation must be extended to spacetime variation at the very least.

What about the right hand side? We all know that mass and volume do not change when rotated. So, mass per unit volume is invariant under rotations alright, but again for sure not under Lorentz transformations. We already know that mass is just a form of energy, and energy is a component of a 4-vector together with momentum. So, generalize mass to a component of a 4-vector.

What about the "per unit volume"? We all learned in school about a rectilinear volume defined by length, width, and height, that is, by spatial extension in the x, y, and z directions. For ease of speaking, let me refer to this as an *xyz* cube. But now Minkowski comes along and promotes 3-dimensional space to 4-dimensional spacetime. We then also have cubes with extension in the time direction, in particular, an *xyt* cube, a *yzt* cube, and a *zxt* cube. So you see that there are four distinct kinds of cubes in 4-dimensional spacetime. Thus, when we speak of "per unit volume" in spacetime, we have four possibilities: the volume could be an *xyt* cube, for example. Is it weird, or is it natural, to have 4 kinds of volume?

All your life you have to deal with the number 3, being the dimension of the space we were all born into. But by now you have been sensitized by Minkowski to the number 4, and may even be able to guess that these four types of "per unit volume" are associated with the four components of a 4-vector. So the bottom line is that mass per unit volume, which we can think of as "mass" linking hands with "per unit volume," gets promoted to two 4-vectors linking hands together, namely a mathematical object known as a 4-tensor, known to physicists as the energy momentum tensor.

I spoke of the kindness of Nature to theoretical physicists repeatedly since chapter 1. Here is another example of that kindness. By the end of the 19th century, physicists have long been acquainted with 3-tensors in 3-dimensional space. They pop up all over classical physics, from fluid dynamics to the rotation of rigid bodies. (Mechanical engineers are particularly familiar with tensors because stress and strain in solids are described by a 3-tensor.[49])

So, interim summary: the equation for gravity has been generalized to

> spatial temporal variation of the gravitational potential ϕ = energy momentum tensor

We are far from being done! Now that the right hand side transforms like a tensor, for the equation to make sense the left hand side must also. We have to promote Newton's gravitational potential ϕ to a tensor.

Aha, we have a tensor lying around, the metric tensor $g_{\mu\nu}$. Furthermore, we know from the Newtonian limit of the the Schwarzschild spacetime that the gravitational potential is hiding in the time-time component g_{00} of the metric tensor! So, replace ϕ by $g_{\mu\nu}$, and then modify the left hand side of the schematic form of the equation for gravity to spatial temporal variation of the metric tensor.

> spatial temporal variation of the metric tensor $g_{\mu\nu}$ = energy momentum tensor

And now we must pay respect to Einstein's equivalence principle. Recall that Einstein generalized the Lorentz transformation to general coordinate transformation $x \to x'$, under which the unprimed and primed observers would see different gravitational fields. We know how the energy momentum tensor on the right hand side of this equation transforms, and so the left hand side must transform the same way. This requirement, known as general coordinate covariance, imposes an extremely stringent condition of what the left hand side must be.

Incidentally, covariance is a general requirement in physics, that the two sides of an equation must change together, that is, change in the same way. (I may be beating a dead horse here, but once again we have the exact opposite of the "truth is relative" doctrine. The unprimed and primed observers see different energy momentum tensors and different gravitational fields, but they must see the same equation! The horse may be dead, but still I sigh. Such a silly misconception, and about something that Galileo understood. Two observers moving relatively to each other and watching a flying baseball would record different velocities of the baseball, but yet they must agree that the same physics governs the flight of the baseball.)

Physics must not depend on the physicist.

Modifying the left hand side, namely the spatial temporal variation of the metric tensor, so that it transforms just like the right hand side, is an extremely challenging task even for someone as off-scale intelligent as Einstein. The historical irony, as I have already alluded to, is that Riemann already solved this problem for space, and it is easy to extend his work to spacetime. The expression that Einstein spent years searching for turns out to be, in hindsight, a fairly simple mathematical object constructed out of the Riemann curvature tensor. The final answer is, in schematic form (compare with the schematic form of Newtonian gravity given earlier),

Einsteinian gravity
curvature of spacetime = energy momentum tensor

and in mathematical form

$$R^{\mu\nu} - \frac{1}{2}g^{\mu\nu}R = 8\pi \, GT^{\mu\nu}$$

Trumpets! Behold the wondrous glorious stupendous tremendous Einstein's field equation!

Figure 9. Einstein writing his famous equation for gravity, in a photo that appears on the cover of the Japanese translation of my book *An Old Man's Toy*. The book designer must have decided that Einstein was left handed! Surely, he or she is acquainted with the letters R and k, but people are so awed by Einstein and curved spacetime that they figure that these strange backward letters must be some secret symbols that physicists use. I don't know what the symbol after 0 means. It looks like a question mark to me. Reproduced from TBS-Britannica.

A parade of greats, besides Einstein, brought us to this, one of the most profound statements in physics: The distribution of energy in spacetime governs the curvature of spacetime. Here G denotes Newton's gravitational constant, $R^{\mu\nu}$ and R the Ricci curvature and the scalar curvature respectively, expressions constructed out of the Riemann curvature tensor. I will let you guess why physicists use the letter R on the left hand side of this equation.

I can't resist showing you Einstein writing his famous field equation in the absence of an energy momentum tensor, that is with $T^{\mu\nu} = 0$. In this case, the scalar curvature R also vanishes. Einstein's equation simplifies to

$$R^{\mu\nu} = 0$$

See figure 9. One further unimportant detail: instead of using the Greek letters μ and ν for superscript, more or less the standard now, Einstein used the Latin letters i and k for subscript.

A final executive summary of Einstein gravity in three "easy" steps:

Gravity, like everybody else, has to live in spacetime.
From everyday life, Einstein was able to abstract the equivalence principle.
Einstein was led to curved spacetime and thence to endow it with dynamics.

Einstein started out unifying space and time into spacetime, and then, trying to accommodate gravity within spacetime, was compelled to curve spacetime. By curving spacetime, he effectively gave life to spacetime and made it dance, warp, and wave.

All the stuff is easy to understand; just think of the number of physics majors that all nations of the world turn out every year. I want to encourage the reader. Surely, if most of them can understand the stuff, so can you.

Notes

[1] "On the Electrodynamics of Moving Bodies," *Annalen der Physik*, 17, 1905, reprinted in *The Principle of Relativity*, Dover Publications, 1952. In this reprinted version, Einstein said explicitly that in 1905 he was not aware of Lorentz's paper.

[2] In German, *Einsteinsche Relativitätstheorie.*

[3] I tried extremely hard to avoid the term "general relativity" in my textbook *Einstein Gravity in a Nutshell.*

[4] Einstein lived at a time when rapid train travel became universal in western Europe, thus making it necessary to synchronize time between different towns.

[5] It is rather obvious, but we can also attach some verbiage to make the obvious more obvious. See *GNut*, pages 159–160.

[6] I, and others in my world, like to think of Einstein as the exterminator of relativity. See, for example, *G*, page 51, and *QFT ASAP*, page 50.

[7] I read that Einstein was even more irritated by certain art historians who proclaimed him "the father of cubism."

[8] Long ago, I was invited to speak at a philosophy conference. Toward the end of my talk, somebody in the audience shouted, "You are a Platonist!" and soon the audience degenerated into two groups, with some shouting, "No no, he is an Aristotelian!" The truth is, I have never heard of a speaker at a physics conference being accused of being a Heisenbergian or a Schrödingerist.

[9] A. Einstein, *The Meaning of Relativity*, Princeton University Press, page 2.

[10] The other was being shown a compass as a child of four or five by his father.

[11] From Paul Arthur Schilpp, ed. *Albert Einstein: Philosopher-Scientist* (Harper Collins).

[12] Well,

$$t'^2 - x'^2 = p^2(t + qx)^2 - r^2(x + ut)^2 = (p^2 - r^2 u^2)t^2 + 2(p^2 q - r^2 u)xt + (p^2 q^2 - r^2)x^2$$

Requiring this to be equal to $t^2 - x^2$ gives us three equations for three unknowns: $p^2 - r^2 u^2 = 1$, $p^2 q = r^2 u$, and $p^2 q^2 - r^2 = -1$. Now your knowledge of high school algebra gets to shine! (Notice that there isn't any quadratic equation to solve, since p, r do not appear, only p^2, r^2.) The solution is $p^2 = r^2 = 1/(1 - u^2)$ and $q = u$. For a more elegant derivation, see endnote 14.

[13] Interestingly, in 1887 the German physicist W. Voigt came close to having this transformation. In Voigt's transformation the right hand side was divided by an extra factor of $\sqrt{1 - \frac{u^2}{c^2}}$. Not knowing Voigt's work, in 1895 Lorentz derived the transformation in a better form than Voigt's. J. Larmor found the exact form in 1900. Not knowing Larmor's work, Lorentz discovered the exact form in 1904. In 1905, H. Poincaré, knowing only of Lorentz's work, developed the transformation further and named it the Lorentz transformation. As for Einstein, he only knew the 1895 version of the Lorentz transformation. The term Lorentz transformation is an example of the Matthew principle: Whoever has will be given more, ... Whoever does not have, even what he has will be taken from him (Matthew 13:12).

[14] Here is a mathematically slightly more advanced derivation, but not beyond that of a bright high school student. Introduce the so-called light cone coordinates $x^+ = (t + x)$, $x^- = (t - x)$, and observe that the invariant $t^2 - x^2 = (t + x)(t - x) = x^+ x^-$. Evidently, $x^+ x^-$ is left invariant if we multiply x^+ by some factor and divide x^- by the same factor. The Lorentz transformation falls out almost instantly:

$$x'^+ = e^\varphi x^+ \quad \text{and} \quad x'^- = e^{-\varphi} x^-$$

with φ some real parameter called the boost angle in particle physics.

To relate this form to that given in the text, simply add and subtract: $t' = \frac{1}{2}(x'^+ + x'^-) = \frac{1}{2}(e^\varphi x^+ + e^{-\varphi} x^-) = (\cosh \varphi) t + (\sinh \varphi) x$, and similarly $x' = \frac{1}{2}(x'^+ - x'^-) = (\sinh \varphi) t + (\cosh \varphi) x$. It is easy to relate the boost parameter or angle φ to the relative velocity u. From the condition that $x' = 0$ implies $x = -ut$ we discover that $u = \frac{\sinh \varphi}{\cosh \varphi} = \tanh \varphi$. (Purists might frown at physicists calling φ an angle since it ranges from $-\infty$ to $+\infty$ as u ranges from -1 to $+1$, but the terminology has the virtue of emphasizing the connection with the rotation angle.) Using the identity $\cosh^2 \varphi - \sinh^2 \varphi = 1$ we then obtain $\cosh \varphi = \frac{1}{\sqrt{1 - u^2}}$ and $\sinh \varphi = \frac{u}{\sqrt{1 - u^2}}$ and hence the Lorentz transformation given in the text.

[15] See, for example, *FbN*, chapter 1.

[16] In fact, if we simply postulate that the universe has a speed limit, and if we amend the Galilean law for the addition of velocities to accommodate this speed limit, we will arrive at this law for the addition of velocities, and once we have this law we could even reverse engineer to derive the Lorentz transformation.

[17] Writing down something like $Kdt^2 + dx^2 + dy^2 + dz^2$ doesn't do it since we could simply absorb $\sqrt{K}$ into the definition of t.

[18] Tragically, within four months of these stirring words, Minkowski, at the age of forty-four, would allegedly express on his deathbed his regrets of not being able to participate in the further development of the concept of spacetime. He had turned Einstein's thoughts toward geometry, and the two seemed destined to collaborate. If that had happened, with Minkowski's mathematical strengths, Einstein gravity would almost surely have arrived earlier than 1915.

[19] Newton was not the only one with a competent faculty of thinking (see chapter 4). The Marquis Pierre-Simon de Laplace had the foresight to speculate about the speed of propagation

c_G of the effect of gravity. Not only that, he was also among those who believed that light moves at some finite speed c. Laplace thought that if gravity was due to some tiny particles zipping back and forth (I am impressed that Laplace's picture is eerily similar to the modern quantum field theoretic view of the graviton zipping back and forth) at the speed c_G, which he supposed (erroneously) to be much larger than c, the time delay could solve some outstanding puzzles about the moon's orbit. We now understand that the speed of propagation is a universal constant, that $c_G = c$ exactly, for the simple reason that the graviton, the quantum of gravity, and the photon, the quantum of light, both propagate in spacetime. The speed of propagation is a property of spacetime, rather than of gravity or of electromagnetism. (But before this understanding, it would seem strange, perhaps even bizarre, that gravity waves and electromagnetic waves would propagate at precisely the same speed c, as some journalists noted.) Conceivably, some bright young guy in another civilization far far away could have proposed the existence of a gravity wave propagating at the speed of light long before a complete understanding of curved spacetime was established.

[20] I haven't proved this to you, as it requires a bit of digression.

[21] The reason for this is that there is only one time in physics, but the number of spatial coordinates could be whatever one's little heart desires. Two timing theories are notoriously difficult to make sense of. If we call t the fourth coordinate as was done in the early days, then every time somebody adds a spatial coordinate the numbering for time would have to be changed.

[22] Or something more involved known as a tensor. This caveat is needed for the electromagnetic field, for example .

[23] A quicker way of seeing this is to note that we could also write a Lorentz transformation with Δt, Δx on the left hand side and $\Delta t'$, $\Delta x'$ on the right hand side, by changing u to $-u$. Time dilation depends only on u^2, not on u.

[24] None of these so-called paradoxes is paradoxical at all. See, for example, *GNut*.

[25] For a dramatized version of this astonishing loss, see *GNut*, pages 7–9.

[26] Historically, Dirac did not realize that he had made the momentous discovery of antimatter. Rather, he naturally sought to identify this unexpected positively charged particle with the only positively charged particle known at that time, namely the proton, which is about two thousand times more massive than the electron, and failed.

[27] The plural is used here because particles typically carry charges other than electric!

[28] For instance, when I talk about vacuum polarization, the process mentioned in chapter 4 in which a photon turns itself into an electron positron pair, which then annihilate each other a moment later producing a photon that continues on its merry way, some students might ask, "Aren't the electron and the positron speeding away from each other? How do they turn back and meet each other again?" In the corresponding Feynman diagram, the electron and the positron are indeed depicted by curved lines. So the answer is: these lines are actually waves.

[29] *QFT Nut*, page 113.

[30] Whether or not this energy could be extracted is outside the purview of spacetime geometry and requires conquering another area of physics, namely nuclear physics, as is well known.

[31] Einstein's derivation in this brief note, in the glare of hindsight, was unnecessarily complicated. Later in 1946, in a lecture at the Technion in Haifa, Israel, he gave an elegant derivation which, surprisingly, is omitted from most textbooks and so is in danger of being forgotten. A notable exception is the textbook by Baierlein. I am grateful to R. Baierlein for providing me with the original reference: A. Einstein, *Technion Yearbook* 5 (1946), 16. I gave my version of it in *GNut*, pages 232–233. It was also odd that Einstein did not publish this derivation in some widely circulated journal; perhaps all was still chaos in 1946. More likely, Einstein did not think it worthwhile to publish a new derivation of an old result.

[32] Incidentally, many physics students, and not a few physicists, have come to grief by not paying attention to Einstein's caveat "sufficiently small."

[33] Necessarily, I have to hide all kinds of subtleties; for a full understanding, you still have to work through a standard textbook.

[34] A detail that is irrelevant for our purposes here is that they differ from the geographers in their definition of latitude. In physics, latitude equals $0°$ at the North Pole, $90°$ at the equator, and $180°$ at the South Pole.

[35] In particular, a space with a horrible looking metric could actually be flat. For instance, somebody could perform an especially nasty coordinate transformation on a flat metric.

[36] Riemann actually even had some confused thoughts that curved spaces might have something to do with gravity. Unfortunately, he was too early, not knowing anything about Minkowski spacetime. This is one of my favorite examples of having a great idea before its time.

[37] Nowadays, I would expect a better-than-average student to learn the necessary differential geometry needed for Einstein gravity in a matter of months.

[38] Check out the equations for $a(t)$ on *FbN*, pages 243–244, and see if you can solve them!

[39] See, for example, *FbN*, page 240.

[40] For a relatively accessible discussion, see chapter VIII.2 of *GNut*.

[41] The derivation is relatively simple. See *GNut*, chapter VI.3.

[42] Tragically, Schwarzschild died a year later of a painful autoimmune disease triggered on the battlefield. I am fond of telling the students in my course on Einstein gravity that if Schwarzschild could find this solution with heavy artillery fire going on all around him (as he described in his letter to Einstein), they should be able to obtain it with ease in the quiet comfort of their homes.

[43] For true understanding, there is no alternative to reading a textbook, such as my *GNut*, part VII.

[44] I believe that I first heard this mysterious remark in a lecture by Julian Schwinger.

[45] Yes, I know that some prefer to call it gravitational wave. I gave my reasons for preferring gravity wave on page 5 of *G*.

[46] See *FbN*, chapter VII.4.

[47] And even easier using the effective field theory approach mentioned in chapter 4.

[48] See, for example, chapter IX.5 in *GNut*.

[49] For essentially the same reason outlined earlier: stress is force per unit area, so under rotations it behaves like two 3-vectors linked together.

7
UNITY OF FORCES IN THE UNIVERSE

> The effort to understand the universe is one of the very few
> things which lifts human life a little above the level of farce
> and gives it some of the grace of tragedy.[1]

A myriad of physical laws, not!

When we first came down from the trees, we might have thought that there must be thousands, if not millions, of forces in the world. A truly remarkable achievement of physics, triumphantly arrived at through centuries of painstakingly cumulative efforts, not to mention all those episodes of getting stuck in blind alleys, is that the apparent myriad of mysterious forces in the physical universe could be reduced to four fundamental interactions.[2] Ladies and gents, only four!

The number four already blows my mind, but Einstein and others want to reduce four to one.

In our everyday world, forces require contact (as was mentioned in chapter 4). But in the quantum world, with particles moving about as clouds of probability amplitudes (see chapter 3), this notion of contact becomes largely inapplicable and irrelevant. Thus, instead of speaking of forces exerted by one particle on another, fundamental physicists prefer to speak of interactions (as we have been doing in this book): when particles come into the vicinity of each other, they interact, that is, influence each other, via one or more of the four fundamental interactions.

Here is a capsule-in-the-cheek summary of what they do.

G: Gravity keeps us grounded. No way you could be a better person if you float up into the vast and eternal emptiness of interstellar space.

E: Electromagnetism accounts for just about everything you experience in your everyday world. It holds atoms together, governs the propagation of light and radio waves, causes chemical reactions, undergirds biology, and last but not least, stops us from commingling our bodies.

S: The strong interaction causes the sun to provide us light and energy free of charge.

W: The weak interaction stops the sun from blowing up in our face.[3]

Only four interactions to learn about: could physics be any simpler, as per Einstein

Just imagine what it would be like if physics textbooks were weighed down by thousands, instead of just a handful, of interactions. For one thing, fewer students would be attracted to physics as a lifetime pursuit. I wouldn't.

Take for example sound and light.[4] In everyday life, they sure appear to be totally unrelated. Sing while hiking in the mountains, and an echo comes back after a few moments. Light, in contrast, seems to be instantaneous; indeed, even to say that light is something (whatever it is!) that travels from here to there is already a highly nonintuitive and nontrivial statement. Sound could propagate around corners, while light can't (hence the proverbial line of sight.) We can talk but we can't emit light, a fact which could surely be turned into a joke.

Sound is a compressional wave traveling through air, generating regions in time and in space with higher and lower than average air density. The properties of sound, governed by the compressibility of air, can all be traced to the mutual repulsion between air molecules, which is in turn due to the residual effect of the electric attraction and repulsion between the electrons and the protons making up the air molecules. Light, on the other hand, is understood to be an electromagnetic wave consisting of electric and magnetic fields varying in spacetime. But repulsion between air molecules and electric and magnetic fields generating each other are both governed by Maxwell's electromagnetic equations. Sound and light were revealed to be different manifestations of one single interaction.

That physics at its most fundamental level drives toward simplicity and unification is not a matter of opinion or preference, but a reality shown to us empirically.

Students failing elementary physics are overwhelmed by a truckload of laws,[5] but most of these are just so-called phenomenological laws, superficial manifestations of the underlying laws.[6] For example, Hooke's law tells us that the amount a spring stretches is given by a constant characteristic of that particular spring times the applied force. This simple linear relation holds only if the stretched spring is far from the breaking point, of course. Hooke's law exemplifies what I call the dull function hypothesis,[7] that most of the functions in elementary physics are pretty boring.

Reduction and unification

The themes of this chapter, reduction and unification, are evidently intertwined with the theme of chapter 1, that the world is comprehensible.

The physics that was so assiduously understood over the centuries could now be mastered in four years of undergraduate study followed by a few years of graduate school. That this is even possible is in itself almost miraculous and speaks to the sweeping unity and utter simplicity of physics.* Of course, the subject is distilled continuously by textbook writers weeding out the inessential.

And most of what is known is inessential. What is hyped today on the internet as a stunning discovery, even if it turns out to be true, may rate only a footnote, probably much less, in future textbooks. This is even without paying the slightest attention to some of the nonsense (Einstein wrong! Foundation of physics shaken!) promulgated in the popular media.

More is different, yes, but physics is a big tent

Unification reflects the reductionist impulse in theoretical physics, the drive toward understanding the world in simpler and simpler terms. Yes, I know about the "more is different" school of thought,[8] advocated forcefully by the Nobel winning condensed matter theorist Phil Anderson. (I played tennis with Phil when I was young and so I certainly had imbibed a dose of the "more is different" philosophy of physics.)

People who love to stir up controversy, some of them outside the physics community, like to set up "more is different" as opposed to reductionism, as if they were inimical to each other. But nobody is claiming that the reductionist approach is capable of explaining all of physics, and anybody who claims that is clearly nuts. A knowledge of quantum mechanics alone hardly enables you to explain the rich and complicated behavior of water in various circumstances. Nor to predict that the solid it forms suddenly when cooled below 0° C would be translucent. Understanding that the proton is composed of quarks will not help elucidate superconductivity, and

---

*I for one would never survive a course on biochemistry. In college, I barely survived chemistry, not only intellectually but literally.

certainly even Gell-Mann would not have claimed that. String theorists are not able to calculate the properties of elementary particles, at least thus far.

Indeed, theoretical physics is neatly arranged into different levels of reality, each enriching and elucidating the next level up, from quantum field theory to quantum mechanics to classical mechanics, up in this context meaning closer to the human condition.

The vitality of physics is nourished precisely by the vastness of the tent, capable of accommodating all types, from lion tamers to acrobats to clowns, as I mentioned in the prologue. The reduction of acoustics to the collective motion of air molecules and the collision between them, and hence to mechanics, certainly does not imply in the slightest that we would have to deny the beauty of a soprano voice, nor the stirring success of quantum mechanics in accounting for molecular collisions. The controversy, if any still exists, is not about physics, but about competition for funding between different areas of physics (and to some extent about taking up space on the web).

Forces and interactions

To explain how the myriad of forces could be reduced to four fundamental interactions would fill books. Let me cite just one example. After banging our heads and learning that we couldn't walk through walls, we were informed later in school that apparently solid everyday objects actually consist of largely empty space[†] sprinkled here and there with quantum blobs called atoms and molecules. How does the apparent solidity of the world arise? Well, given the large variety of solids, the theory of solids can get mighty complicated. But a simple cartoon picture suffices here: the nuclei of the atoms comprising the solid are locked in a regular lattice, while the electrons cruise between them as a quantum cloud. A collective society in which all individuality is lost! The electrons no longer exist as separate entities. The arrangement is highly favorable energetically; that is jargon for saying that enormous energy is required to disturb that arrangement. Revolution is costly. It takes quite a tough guy to crack a rock into halves.

What we see in everyday life is by and large due to some residual effect of the electromagnetic force: since common objects are all

---

[†]When I first read about this fact, I was astonished. Is there anybody who wasn't?

electrically neutral, consisting of equal numbers of protons and electrons, the electromagnetic force between these objects almost all cancels out. Even the steel blade of a jackhammer smashing into rock is but a pale shadow of the real strength of the electromagnetic force. Just about the only time the true fury of electromagnetism shakes us is when thunder and lightning fill the sky. While we have totally conquered electromagnetism, all ancient people attribute its occasional bursts of temper to the gods.[‡]

Trying to narrate the drive toward unification in physics in a single chapter, my goal can only be to give you an overview of this drive, of the layout of the forest with barely any mention of the trees. Those readers interested in the branches, or perhaps even the varieties of bark, are referred to the many books available.[9]

Range versus strength

While the jolly crowds enjoy the use of gravity and electromagnetism continuously without thinking much about it, they feel no personal ties to the strong and the weak interactions simply because these two interactions are short ranged. Indeed, operating only on the nuclear and subnuclear scales, the two interactions were not even known to physicists till the turn of the 20th century. The strong attraction between two protons falls abruptly to zero almost as soon as they move away from each other. The weak interaction has an even shorter range, roughly a thousand times shorter than that of the strong interaction. That the strong and weak interactions are short ranged also implies that they cannot support waves propagating over macroscopic distances. For this reason, they did not come into physics, let alone the general consciousness, till the end of the 19th century, millennia later than the other two interactions.

In contrast, the gravitational force between two masses and the electric force between two charges both fall off with the separation R between the two objects like $1/R^2$, the inverse square law celebrated in song and dance. For R large, these forces still go to zero, but slowly enough that we can "feel" the tug of the sun, literally an astronomical distance away.[10] For that matter, our entire galaxy, the Milky Way, is falling toward our neighbor the Andromeda galaxy.

[‡]We still devote one day a week to electromagnetism: Thursday is Thor's day.

Gravity and electromagnetism are known as long ranged and thus can and do support propagating waves. We have seen electromagnetic and gravity waves[11] since 1886 and 2015 respectively.[12]

Thus, in the contest between the four interactions, brute strength is not the only thing that counts: many phenomena depend on an interplay between range and strength. An important example is fusion versus fission in producing nuclear energy.[13] In a large atomic nucleus, famously the uranium nucleus, the electric repulsion wins over the strong attraction. Each proton is strongly attracted by only the protons or neutrons right next to it, but it feels the long ranged electric repulsion from all the other protons in the nucleus. The repulsion wins and the nucleus wants to split into two smaller pieces, accompanied by the release of energy. In contrast, when two small nuclei get together, each consisting of a few protons and some neutrons, the strong attraction overwhelms the electric repulsion and they want to fuse into a single nucleus.

The huge advantage of fusion over fission, on which the human race rests its hopes, is of course the inexhaustible supply of hydrogen atoms in water compared to the rare uranium bearing ores. Essentially all of the energy we consume, from the food we eat to the gasoline we burn, comes ultimately from nuclear fusion in the solar interior.

Einstein tried to marry the wrong pair

> Let no man join together what God has put asunder.
> —Wolfgang Pauli, laughing at Einstein's futile labors[14]
> trying to unify electromagnetism and gravity

In contrast to the first half of Einstein's career as a theoretical physicist, in which one brilliant triumph followed another, the second half was doomed[§] by his futile search for a so-called unified field theory.[15] Perhaps because the strong and the weak interactions were not identified and distinguished till long after Einstein's student days, he ignored them, expecting the two to pop out as epiphenomena once he unified the electromagnetic and the gravitational interactions.

[§] Even so, he contributed brilliantly to the quantum physics of entanglement (as I recounted in chapter 3; spooky!), which we humans are still desperately trying to harness for our own benefit.

Unifying the various interactions could be likened to doing a jigsaw puzzle. The pieces representing electricity and magnetism were clearly meant for each other and by the late 19th century were already linked together into a larger piece. Einstein thought that this piece would fit the piece representing gravity, but just couldn't mash them together.

Einstein did not live to watch his drive toward unification taking an unanticipated turn.

Surprise surprise! In the late 1960s and early 1970s, the venerable grande dame, the long ranged electromagnetic interaction, was unified with the scraggly latecomer to the party, the shortest ranged, indeed, the almost zero ranged, weak interaction, into what became known as the electroweak interaction, as I have already alluded to in chapter 1. The two oldest interactions known to physics, gravity and electromagnetism, spurned what on the surface looked like a match made in heaven: both long ranged, both involved in the tides that drive the affairs of men and women, both well understood. But no, electromagnetism had to choose the almost deranged weak interaction[16] with all its bizarre idiosyncrasies.¶ Such is the mystery of "mutual attraction."

Einstein had the last laugh on Pauli

Soon after the electromagnetic and the weak interactions were merged into the electroweak interaction, the electroweak interaction was in turn merged with the strong interaction into what became known as the grand unified interaction, about which much more presently. (There are of course always negativists who doubt, and we will come back to them, but for the moment let us proceed as if grand unification were established if only for narrative continuity.)

But yet, in a real sense, Einstein had the last laugh on Pauli. I would argue that grand unification realizes Einstein's impossible quest. Physicists have joined together what God has only appeared to put asunder. While it is true that unification of the other three interactions, leaving gravity out, was quite different from what Einstein had in mind, his vision of a unified design inspired the grand unifiers of the 1970s and continues to inspire us today. With the strong, weak, and electromagnetic interactions grand unified into a gauge theory generalizing electromagnetism, the search for

---

¶The Nobel laureate Yoichiro Nambu referred to the weak interaction as "God's mistake."

unification with gravity once again amounts to the unification of two geometric theories, as Einstein had wanted. In this ironic twist of history, Einstein turns out to be right in spirit, if not in detail.

Hopeless befuddlement led to the weird word "gauge"

Warning: This section turns out to be rather dense, but fear not, I will give a one sentence summary at the end. I have to include this section if only to explain the curious phenomenon, occasionally overheard, of a bunch of chattering theoretical physicists using the uncommon word "gauge."

Yes, but what does gauge in gauge theories have to do with railroad gauges or gauged skirts?

To explain this (even in the barest terms), we have to go back to Heisenberg rotating the proton and the neutron into each other back in chapter 5. For the laws governing the strong interaction to remain unchanged, the angles of rotation must not depend on where we are in spacetime. In other words, we have to rotate the two nucleons by the same amount everywhere in the universe. In physics talk, this is known as a global transformation, generating a global symmetry.** Recall also that with the advent of quarks, this global transformation became understood as a rotation of the up and down quarks into each other.

Some years earlier, back in 1918, the mathematical physicist Hermann Weyl proposed that the laws governing electromagnetism would remain unchanged upon multiplying the electron wave function[17] by a factor that could vary according to where you are in spacetime, in contrast to a global transformation. Weyl's transformation naturally became known as a local transformation.

In comparing Weyl with Heisenberg, you might wonder, quite aside from this business of local versus global, why Weyl was not rotating the electron into anything. That was simply because there was nothing else (before the discovery of the neutrino) for the electron to be rotated into, in contrast to the proton and the neutron occurring naturally as a pair (if we ignore the proton's charge.)

Actually, Weyl was hopelessly befuddled, and what he wanted makes no sense whatsoever in the glare of hindsight. I am telling this story

---

** Just from this terminology, you could see that, before our attention deficit age in which everything has to be super and ultimate, physicists were much more modest, equating "everywhere" with the known world, aka the globe.

partly in order to explain where the now ubiquitous word "gauge" came from. The transformation Weyl had in mind involves changing the distance between points: he was somehow inspired by Einstein's work on gravity. Hence the term "gauge" noted earlier about railroads and skirts.[18] So Weyl named this kind of local transformation a gauge transformation, and the corresponding theory a gauge theory.

Nowadays, every undergraduate who has had a course on quantum physics knows that Weyl should be multiplying the electron wave function by a complex number characterized by a phase angle.[19] But we should excuse Weyl for not realizing that complex numbers were necessary since he was doing this before quantum physics had been written down in the form as we now know it. But henceforth, for better or for worse, this peculiar word "gauge" became one of the most used words among some theoretical physicists.

Among the cognoscenti, electromagnetism is sometimes referred to as an abelian gauge theory, as was already mentioned in chapters 1 and 5. The only take home message you need to know is that an angle is involved, and so the invariance group of electromagnetism turns out to be the group $SO(2)$, the invariance group of the circle, in fact the simplest group[††] one could imagine. In a strange way, there is almost poetic justice that the interaction which we are most familiar with and which underlies modern life (streaming your favorite movies!) is associated with the simplest imaginable group. This is what I mean by Nature's kindness to physicists!

The group $SO(2)$ is essentially the same[20] as the group $U(1)$. In an interesting twist of pedagogy, undergraduate physics majors are by and large unaware of this $U(1)$ group. The reason is that while some manifestation of this symmetry was already known in the late 19th century it did not play much of a role in classical electrodynamics. In fact, the eminent theoretical physicist Oliver Heaviside thundered at the time that the mathematical machinery associated with this symmetry should be stuffed "into the dustbins of history." He was not wrong in thinking thus, because he had no way of knowing[21] about quantum physics. But with the rise of quantum electrodynamics, this group $U(1)$ ended up playing a central role. As was mentioned just now, the symmetry transformation corresponds to

---

[††]I use the word "group" throughout in order not to confuse the casual reader, but strictly speaking, I should refer to them as Lie algebras.

multiplying the electron field by a complex phase, words that would have amounted to gibberish to 19th century physicists. Furthermore, nobody thought of the electron as a field till the 1930s.

Executive summary as promised: Heisenberg rotates the proton and the neutron fields into each other through angles that are not allowed to depend on where we are in spacetime, while Weyl (with some after the fact corrections) multiplies the electron field by a phase angle that depends on where we are in spacetime.

Combining Heisenberg and Weyl brings in the gauge bosons

The next momentous development came when in 1954 Chen-ning Yang and Robert Mills, and others, proposed combining the notions advanced by Heisenberg and Weyl. Well, can you do it in the glare of hindsight?

Yes! Promote the global $SU(2)$ symmetry of Heisenberg to a local symmetry[‡‡] by allowing the angles through which we rotate the two nucleons into each other to depend on where we are in the universe. To say this in another way: with electromagnetism switched off we are free to rename any linear combination[22] of the two nucleons the proton, and the other combination the neutron, and that this "labeling" could change continuously from one point to the other in spacetime.

This was the crucial step that allows the electromagnetic and the weak interactions to embrace each other to form the electroweak interaction. In technical language, the $U(1)$ gauge group of electromagnetism was promoted to the group $SU(2) \otimes U(1)$. I won't be able to explain here[23] what the multiplication symbol with a circle around it means, except to opine that there are many kinds of multiplication on heaven and earth that Horatio had never dreamed of.[24]

A defining feature of Yang-Mills theory (also known as nonabelian gauge theory) is the presence of massless particles known as gauge bosons. Roughly speaking, they are required to communicate the fact that the symmetry transformation varies from point to point in spacetime. The number of gauge bosons is determined by the group the theory is based on (and carved in stone as I emphasized in chapter 5). For $SU(n)$, with n greater

---

[‡‡]At the risk of confusing the reader, modern field theory textbooks explain that, strictly speaking, a local symmetry is not a symmetry, but a redundancy of description. See, for example, *QFT Nut*, page 189.

than or equal to 2, the number is[25] $n^2 - 1$. Thus, for $SU(2)$, the first nonabelian gauge theory discussed by Yang and Mills, the number of gauge bosons equals $2^2 - 1 = 4 - 1 = 3$.

We want our universe to be illuminated

For electroweak unification based on $SU(2) \otimes U(1)$, the number of gauge bosons is four, one of which is the familiar photon.

A deep result in quantum field theory indicates that these massless bosons would generate long range forces.[26] But in the real world, leaving gravity aside and focusing on the electromagnetic and the weak interactions, we know only one long range force, namely the one we live with everyday, the one generated by the photon. These three other gauge bosons (now known[27] as W^+, W^-, and Z^0) have to be made massive. Easier said than done! The way to do this, called the Higgs mechanism,[28] earned Franqis Englert and Peter Higgs the Nobel prize.[29]

It is crucial that all these machinations leave one and only one gauge boson massless, to be identified as the photon. Were a theorist to inadvertently render all four gauge bosons massive, he or she would be stuck with a universe, oops, without any light.

In modern physics, the cry "Let there be light!" has been replaced by "Let one, but only one, of the four gauge bosons in electroweak unification remain massless!" Much less majestic, but closer to the truth.

With these few sentences I am glossing over about two decades of arduous and puzzling work[30] by both theorists and experimentalists. In the next section, I will give a bit more detail about electroweak unification, but just barely enough so that we could talk about the even more intricate grand unification.

Electroweak unification

First, after the advent of quarks in the 1960s, we talk about the up quark and the down quark instead of the proton and the neutron. For our purposes here, all you need to know about quarks is that the proton consists of two up quarks and one down quark: $p = (uud)$, while the neutron consists of one up quark and two down quarks, $n = (udd)$. Proton is mostly up while neutron is mostly down, so to speak.

The electroweak interaction resulting from the marriage of the electromagnetic and the weak interactions may be summarized by writing

$$\begin{pmatrix} u \\ d \end{pmatrix} \quad \bigg| \quad \begin{pmatrix} v \\ e^- \end{pmatrix}$$

The superscript in e^- reminds us that the electron carries negative[31] electric charge.

Think of two houses in a fairy tale, each with two stories. In one house the down quark d lives downstairs while the up quark u lives upstairs. (Recall how Gell-Mann came up with the whimsical names up and down[32] quarks.) In the other house, the electron e^- lives downstairs while the neutrino v lives upstairs. (Again, for our purposes here, all you need to know about the neutrino,[§§] denoted by the Greek letter v, is that it is a mysterious particle[33] carrying no electric charge and which participates only in the weak interaction.[¶¶] Hence the neutrino was for a long time hidden from physicists.)

The two houses are separated by a fence, indicated by the vertical line. On one side of the fence, the quarks participate in the strong interaction, while on the other side the neutrino and the electron do not. Particles that do not participate in the strong interaction are known as leptons. (The lepton is the least valuable[34] coin in ancient Greece, and a person with a thin narrow face is leptoprosopic.[35])

Thus far, we have not talked about the four gauge bosons responsible for the electroweak interaction. The photon we have known since birth, so we know what it is capable of. What about the other three gauge bosons, W^+, W^-, and Z^0, which Higgs had rendered massive through his mechanism? For our purposes here, I will ignore the Z^0. I will now show you how readily you could become acquainted with weak interaction processes.

Think of the W^+ boson lifting the electron upstairs to become a neutrino, and the down quark upstairs to become an up quark. This could be written as[36]

$$W^+ + e^- \leftrightarrow v \quad \text{and} \quad W^+ + d \leftrightarrow u$$

The left right arrow $\leftrightarrow$ simply reminds us that the process could go both ways.

How about the W^- boson? Yes, you guessed it, the W^- boson acts oppositely to the W^+ boson. It carries the neutrino and the up quark downstairs

---

[§§] Meaning "small neutron" in Italian, and thus named by Enrico Fermi in order to distinguish it from the neutron.

[¶¶] And also gravity of course, like everybody else in spacetime.

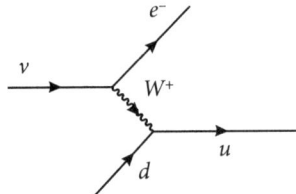

Figure 1. A neutrino turns itself into an electron by emitting a W boson, subsequently absorbed by a down quark, which becomes an up quark. This Feynman diagram is simpler than the ones used to calculate the electron's magnetic moment in chapter 4. Believe me, particle physics is not that hard.

to become an electron and a down quark respectively, written as

$$W^- + \nu \leftrightarrow e^- \quad \text{and} \quad W^- + u \leftrightarrow d$$

The four fundamental processes[37] exemplify a property of quantum field theory known as crossing symmetry. Compare $W^+ + e^- \leftrightarrow \nu$ and $W^- + \nu \leftrightarrow e^-$. You see that if in the former process you move the W^+ from the left hand side of the double arrow to the right hand side and flip its electric charge from $+$ to $-$ (more generally, flipping a particle into its antiparticle), you would obtain $e^- \leftrightarrow W^- + \nu$. But this is exactly the same as the latter process $W^- + \nu \leftrightarrow e^-$ since the double arrow works both ways. Crossing allows us to move a particle from one side of the arrow to the other side as long as we remember to change particle to antiparticle. (Junior high school algebra, anybody?)

Now let me get you started on the road to becoming a high energy theorist. Reading $\nu \leftrightarrow W^+ + e^-$ from left to right, we see the neutrino emitting a W^+ and becoming an electron. Suppose that a down quark d is lurking about in the vicinity. Then reading $W^+ + d \leftrightarrow u$ from left to right, we see the down quark d absorbing the W^+ emitted a moment earlier by the neutrino, thus turning itself into an up quark u. See figure 1.

The net effect is the process $\nu + d \rightarrow e^- + u$, namely, a neutrino colliding with a down quark producing an electron and an up quark, typical of the kind of process high energy experimentalists dreamed to see by begging and cajoling various governments to cough up the money for constructing huge accelerators. The particular experiment described here was actually done in the 1970s, smashing a beam of high energy neutrinos into a junked battleship and watching an electron come flying out. You may not think of junked battleships this way, but they (as any pile of junk for that matter) are rich sources of protons and neutrons, and hence of up and down quarks.

This kind of "game" might remind some readers of high school chemistry except that the "reactions" are occurring in the world of quarks rather than in the world of chemicals. Or, perhaps at an even lower level, of some children's toy that consists of joining together some smaller blocks to form larger blocks. For particle physicists the basic building blocks are fundamental processes such as $W^+ + e^- \leftrightarrow v$.

Referring back to our picture of two houses separated by a wall, we see that charge conservation mandates that if we move the upstairs guy downstairs in one house, say v to e^-, then we must[38] move the downstairs guy in the other house upstairs, thus d to u. Hence[39] $v + d \rightarrow e^- + u$.

By crossing, as you have just learned a moment ago, we could also have the process $d \rightarrow \bar{v} + e^- + u$. (Since the neutrino does not carry charge, the antineutrino is denoted by an overbar.) Furthermore, noting that the right arrow in $v + d \rightarrow e^- + u$ is actually a double arrow that works both ways, we could write this is $e^- + u \rightarrow v + d$ and cross the electron to obtain $u \rightarrow e^+ + v + d$. This describes the up quark going into a down quark emitting a positron and a neutrino.

You see that once you get the hang of this "game," you could proceed to write down all the weak interaction processes studied for much of the 20th century. A doctorate in particle physics might yet be on your horizon.

The strong interaction gluing quarks together

Even though I have rushed over the weak interaction with the greatest of haste, that still leaves me with practically no room for the strong interaction, the other of the two interactions willfully ignored by Einstein. I have already mentioned that the strong interaction operates only on the quarks' side of the fence, not on the neutrino and the electron. I give you here the briefest possible account of the strong interaction.

Gell-Mann, the father of quarks, made the crucial suggestion that quarks such as u and d in fact each come in three different colors, say red, yellow, and blue. We have three kinds of up quarks, u^r, u^y, and u^b in an evident notation. Similarly for the down quark. We could think of the house in which u and d live as actually subdivided into three attached townhouses or condos separated by interior walls. In other words, we subdivide the house the quarks live in into three townhouses:

$$\begin{pmatrix} u \\ d \end{pmatrix} \rightarrow \left(\begin{array}{c|c|c} u^r & u^y & u^b \\ d^r & d^y & d^b \end{array} \right)$$

In the red townhouse, the red up quark u^r lives upstairs and the red down quark d^r lives downstairs. Similarly for the yellow townhouse and for the blue townhouse.

The W^+ and W^- bosons cannot go through the walls (represented by the vertical lines) separating the differently colored townhouses. For example, the W^- can carry the red up quark u^r downstairs to the red down quark d^r, but not to the yellow down quark d^y. The W^+ and W^- bosons cannot change the color of the quark they carry up and down the stairs.

Fine, all this carrying quarks up and down stairs generates the weak interaction we observe in the universe. But what about the strong interaction? Once again, we invoke Yang-Mills theory, with a local symmetry group $SU(3)$ transforming the three colors into each other.

Each set of gauge bosons is capable of its own suite of wizardry

Recall that a gauge theory based on $SU(n)$ has $n^2 - 1$ gauge bosons. Thus, the strong interaction is generated by $8 = 3^2 - 1$ massless gauge bosons, known as gluons for reasons we will soon see. These eight gluons are capable of magic tricks the W^+ and W^- bosons know not how to do: the gluons could change the color of the quarks. In our silly analogy, they could zip right through the walls separating the three differently colored townhouses. For instance, a gluon can transform a red up quark u^r into a yellow up quark u^y, or a yellow down quark d^y into a blue down quark d^b. But in contrast to the W^+ and W^-, gluons cannot carry quarks up and down stairs. A gluon cannot transform an up quark into a down quark, or vice versa. Furthermore, the gluons cannot touch the leptons, unlike the W^+ and W^-.

Each set of gauge bosons is capable of its own suite of wizardry. Is this fun? And truth be told, not that difficult to master. Difficult to invent, yes, but to master, no. Generations of students have learned it.

The gluons thus generate a force[40] between differently colored quarks, and glue them together to form protons, neutrons, and their various cousins. (Hence the name gluons bestowed by Gell-Mann.) Incidentally, the protons and neutrons that made up you and me and the rest of the universe are referred to as colorless[41] combinations, an equal mix of red, yellow, and blue. In fact, the proton $p = (uud)$ and the neutron $n = (udd)$, in each case with the right color mix to render the proton and neutron colorless. Keep this in mind for use shortly.

This foray into the strong interaction was certainly, and unreasonably, lightning fast.[42] But apologies aside, let us summarize. Physicists went from being acquainted with one single gauge boson, the ubiquitous photon that we have effectively enslaved for our purposes, to knowing $12 = 1 + 3 + 8$ gauge bosons, which together generate the strong, the weak, and the electromagnetic interactions, as described in this table.

1 photon	electromagnetism	couples to everybody except neutrinos but cannot move anybody
3 weak bosons	weak	able to move quarks and leptons up and down stairs
8 gluons	strong	can zip between townhouses but cannot change floors, nor cross the fence separating quarks and leptons

Potentially thousands of interactions down to three

I started this chapter by saying that we once thought that there must be thousands, if not millions, of forces (or interactions in modern parlance) in the universe. Now we are down to three, the electroweak and the strong interactions, with gravity lurking in the shadow. Leaving gravity aside, we have a gauge theory of the physical world based on $SU(3) \otimes SU(2) \otimes U(1)$, commonly called the standard model. Truly the worst name ever in the history of physics! Even the 123 theory would be better.

At this juncture in our quest, there are two avenues that have been explored.

The temptation is to unify the electroweak and the strong interactions into a single grand unified interaction, as I had already mentioned.

Then there are physicists who doubt that the time is ripe for grand unification as I had also mentioned. They point to the deficiencies of grand unified theory (which I will come to shortly), and theorize instead about what became known as "physics beyond the standard model." The rallying cry is to keep an open mind about eventual unification, which certainly everybody could agree with. Unfortunately, after enormous amount of effort, there has been no coherent theory[43] of physics beyond the standard model,*** apart from possibly a few hints here and there.

It is important for the reader to know that both of these avenues could hardly claim to be the latest: they have been going on for almost half a century and counting. I am among those who believe in some form of ultimate unification, without of course discounting the possibility of "new" physics beyond the standard model, for instance neutrino masses and a multitude of "private" Higgs fields, both topics on which I have published speculations. (The reader does not have to be concerned with these topics in order to follow the rest of this chapter.)

Grand unification: four interactions down to just two?

With three of the four interactions now based on a gauge theory with the group $SU(3) \otimes SU(2) \otimes U(1)$, an attractive possibility, as proposed by H. Georgi and S. Glashow, is to mash these three groups together to form a single grand unified theory based on the group $SU(5)$.

Why 5? Well, $3 + 2 = 5$ with the $U(1)$ group slipping into the crack between $SU(3)$ and $SU(2)$. (Didn't I tell you that particle theory is easy?)

Applying what you "learned" earlier, you know that this $SU(5)$ gauge theory has $5^2 - 1 = 24$ gauge bosons, of which we have already met 12. (Here is the math: $(3^2 - 1) + (2^2 - 1) + 1 = 8 + 3 + 1 = 12$, as indicated by the preceding table.) The remaining 12, known as X bosons, are made extremely massive by the Higgs mechanism, so that they are way too massive to be produced by any conceivable accelerators we could built. But for some fleeting moments after the Big Bang, they remained free from the Higgs mechanism, staying massless and active.

Go back to the jigsaw puzzle analogy used before. It is as if, after much toil, we have assembled two large pieces, one representing the electroweak interaction, the other representing the strong interaction, and to all appearances they would interlock perfectly with each other, if we could somehow find twelve small pieces to fill in some gaps here and there. Examining the box the puzzle came in, we find a secret compartment containing exactly the twelve extra pieces we need. Exciting indeed! And that was how some physicists felt when grand unification was first worked out.

***Experiments have also come up empty. The Large Hadron Collider, which allegedly was designed to discover a whole zoo of supersymmetric particles, failed to find them.

I will tell you presently what these X bosons can do, but meanwhile, shaking the box to make sure we got everything, we find two trays with fifteen slots to put the quarks and the leptons in. The number of slots, 15, is fixed by group theory, as was emphasized in chapter 5. (In case you are wondering, the math[44] works like this: $5 + (5 \times 4)/2 = 5 + 10 = 15$.) Excited, we count the number of quarks and leptons we have to see if they would fit. Now I have to give you the briefest instruction of how quantum field theorists count, less instruction than merely a flavor of the instruction.

Back in chapter 3, I told you that an electron could spin either up or down. In quantum field theory, the electron field possesses a property known as chirality (from the Greek word "cheir" for hand; see the following), corresponding to these two spin states, very roughly speaking. Thus, there are actually two electron fields (or perhaps better to say "two pieces of the electron field") known as left and right handed.[†††] Similarly for the quark fields, each of them comes in two versions, left and right handed.

But there is one huge exception: the mysterious neutrino field is only left handed. We don't know where the right handed neutrino field is! Remember that earlier in this chapter I mentioned that Nambu said that the weak interaction was "God's mistake"? One of the weird features of the weak interaction is that the neutrino is only left handed, thus leading to parity violation, as was already mentioned back in chapter 2.

Ready to count? Electron plus neutrino: $2 + 1 = 3$. Up and down quarks: $2 \times 3 \times 2 = 12$ (2 hands times 3 colors times two species of quarks). Altogether, $3 + 12 = 15$, exactly the right number of lepton and quark fields to fill the 15 slots in those two trays! (Readers who know some group theory might recognize by trays I am really talking about irreducible representations of the group $SU(5)$.) Not only does the counting work, but the slots in those trays are marked with the electric charges and the color (or lack thereof) of the field that would fit in that slot, and all the electric charges and colors work out perfectly.

It is this kind of perfect fit that bolsters the faith of the believers in grand unification. Physicists who do not believe would simply dismiss all these numerical equalities as just lucky coincidences, which I for one find hard to believe. To keep things simple, I have not even mentioned another deep mystery of the universe: the set (variously known as a generation or

---

[†††]The technical names in quantum field theory provide part of the fun in learning it!

a family) of quark and lepton fields we have talked about is triplicated! As far as we know, for the universe to function as it does, only one generation of quarks and leptons (namely u, d, v, e^-) would suffice. Instead, we have three generations, and the other two are all unstable, decaying into the first generation. This mystery is known as the family problem.

Mixing the quarks and the leptons could stir up big trouble

So, what could these twelve new superheroes do that the gauge bosons we already know, namely the photon, the W^+ and W^-, and the gluons, could not do? Any guesses?

Exactly, you might have guessed it. Not only could they pass right through the walls separating the three differently colored townhouses, they could also jump right over the fence separating the quarks from the leptons, that is, the neutrino and the electron.

You could almost see what would happen when we grand unify. These new bosons are going to mix up the quarks and the leptons! Could stir up big trouble!

Previously, the three interactions, the strong, the weak, and the electromagnetic, are separate and operate on different particles. For instance, the electron practically embodies the electromagnetic interaction, while the neutrino, carrying no electric charge, does not listen to the photon. Furthermore, while the quarks are enthralled by the strong interaction, the electron and the neutrino are completely oblivious to it. As a result, the leptons and the quarks go through their comfortable lives driving along in their own lanes, so to speak. Thus, in the process $v + d \rightarrow e^- + u$ discussed earlier, the v turned into an e^- while the down quark d turns into an up quark u. They don't mix: physicists say that lepton number and quark number are separately conserved.[45] For example, in the process $v + d \rightarrow e^- + u$ we start with one lepton and one quark and end up with one lepton and one quark. The number of leptons and the number of quarks in the universe are fixed once and for all and do not change.

The stability of the universe depends on the proton being slightly less massive than the neutron

To see why grand unification would stir up big trouble for the universe, we have to understand why the electron and the proton are stable while almost all the other subnuclear particles decay in a zillionth of a second.

First, according to Einstein's $E = mc^2$, the mass of a decaying particle must exceed the sum of the masses of the decay products. The reason is simple. Consider a particle A decaying into particle B and particle C. If the mass m_A of particle A is less than $m_B + m_C$, then particle A does not contain enough energy to produce particle B and particle C.

This mass constraint and charge conservation suffice to guarantee the stability of the electron because it is the lightest electrically charged particle.

How about the proton? There is an interesting twist to its stability. Consider the neutron, the proton's identical twin except that it carries no electric charge. Upon the discovery of the neutron, physicists expect the proton to be more massive than the neutron, because the electric field around the proton contains energy, which according to Einstein is equivalent to mass. But no, surprise surprise! This entirely reasonable expectation from elementary physics does not hold: the neutron is actually more massive than the proton.

Thus, it is the neutron, rather than the proton, that is unstable. Indeed, a neutron decays[46] into a proton in about 14 minutes, a weak interaction process caused by one of the down quarks in the neutron decaying into an up quark. Indeed, you the budding particle physicist even wrote down the process responsible earlier, namely $d \to \bar{\nu} + e^- + u$. Or, if you prefer, you could write $(udd) \to \bar{\nu} + e^- + (uud)$, which you might recall is how you would represent a neutron and a proton using quarks. This depicts the neutron n decaying into a proton p emitting an electron and an antineutrino.

Thus, the proton is stable simply because it is slightly less massive than the neutron. For decades, this strange fact, that the electrically charged twin is less massive than its otherwise identical twin, was presented as a puzzle[47] for particle physicists to solve. It is still unsolved: the puzzle has been simply shoved off from the level of the nucleons into the level of quarks. The up quark is less massive than the down quark. Why? Since we don't have a deep understanding of the origin of quark masses, we are still clueless.

This puzzling fact, that the proton is slightly less massive than its twin, provides a bedrock of stability for the universe. How so? A major ingredient of the universe is the hydrogen atom, which, as you know, consists of a negatively charged electron orbiting a positively charged proton, bound to each other by electric attraction. Imagine that, instead of the neutron decaying in 14 minutes, it is the proton that decays into the neutron emitting a positron and a neutrino. The orbiting electron would be attracted immediately to its antiparticle, the positron, with the two of them annihilating each other. All the hydrogen atoms in the universe, including those in our bodies, will soon vanish, and our world is no more.

Please realize that this alarming scenario, which fortunately does not occur in reality because for some reason the neutron is more massive than the proton, is what physicists understood for close to half a century before grand unification. Now grand unification is posing a different threat to the stability of the universe.

Poof goes the universe, gone with the proton

By having gauge bosons that could jump over the fence separating the quarks and the leptons, grand unification has effectively brought them all together into one big house. Perhaps you could almost see that this spells serious trouble for the universe! Oh, for sure, we want to live in a stable universe, one that does not disappear in a poof before we could say abracadabra.

This absolute stability of the proton holds only as long as quark number is conserved. With the X bosons brought in by grand unification and capable of transforming quarks into leptons, quark number and lepton number are no longer separately conserved, and so the proton could decay, via the process shown as a cartoon in figure 2. Or, if you prefer, as in high school chemistry, combining $u \rightarrow X + e^+$ and $X + u \rightarrow \bar{d}$ gives us $(uud) \rightarrow (\bar{d}d) + e^+$, that is, $p \rightarrow \pi^0 + e^+$.

To repeat, those twelve extra gauge bosons, by their ability to transform a quark into a lepton, could make the proton go poof, and hence by extension, the universe go poof.

The universe has stuck around for around 10^{10} years, while grand unified theory predicts proton decay. Thus, at first sight, grand unified theory would appear headed straight to the dustbin of history. But remarkably, for another reason,[48] which I cannot go into here, those trouble-making gauge bosons are rendered extremely massive by the Higgs mechanism, as I already mentioned, thus effectively suppressing their strength. Consequently, the lifetime of the proton turns out to be around 10^{31} years, far longer than the known age of the universe. Quantum physics operates on probabilities, and the more massive the gauge boson responsible, the less likely a given process could occur.[‡‡‡] The probability that a given proton would decay within the next year is thus around 10^{-31}. So all is well.

[‡‡‡]This also means that processes driven by the massless photon and graviton occur with almost total abandon, as we know from everyday life.

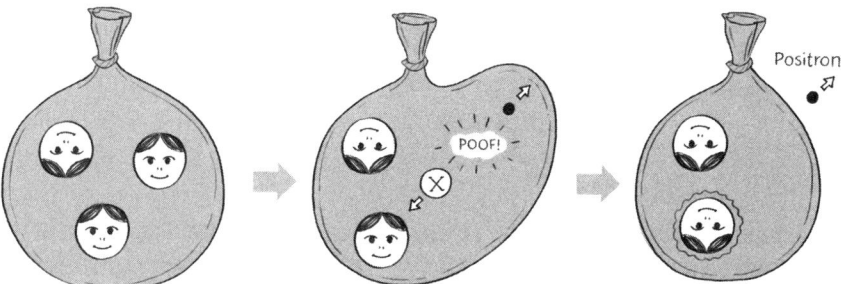

Figure 2. The proton is represented as a bag containing two up quarks and a down quark. One of the two up quarks in the proton emits a gauge boson X, that is, a newcomer gauge boson not in the $SU(3) \otimes SU(2) \otimes U(1)$ theory, and transforms itself into a positron, that is, an antielectron, represented by the black dot. The gauge boson X is absorbed by the other up quark which as a result turns into an anti down quark, represented by the upside down face with a shaded rim to distinguish it from the down quark hanging around all this time as a spectator. The positron escapes, and the anti down quark settles down with the down quark into temporary domestic bliss as a strongly interacting particle called the neutral pi meson π^0. The proton thus decays: $p \rightarrow \pi^0 + e^+$. Redrawn from A. Zee, *Quantum Field Theory, as Simply as Possible*, Princeton University Press.

I am well aware that some nattering nabobs of negativism are constantly harping[§§§] about some difficulties faced by grand unified theory, one being that experimentalists have been watching a lot of protons for a long time and have yet to see any of them decaying.[49] Another objection is that going from 12 gauge bosons to 24 gauge bosons hardly seems moving in the direction of unity and simplicity. But this is silly: it reminds me of people saying that Mendeleev's periodic table with its 92 elements represents a step backward from the 4 or 5 elements proposed by several ancient civilizations.[50] I am among those with faith that, in the ripeness of time, grand unified theory will triumph. The fit between the strong interaction and the electroweak interaction into a single interaction is so seamless that it boggles my mind to think it was all accidental, a mere coincidence. If it does turn out that the fit is illusory, then my theorem about Nature's kindness would be shattered. In my humble opinion, this would render physics a far less worthy subject for contemplation.

I might mention in passing another possible major accomplishment of grand unified theory. Many physicists would prefer, for aesthetic "reasons," to have the universe start out in a pristine state, becoming

---

[§§§]It would be ridiculously easy in the early 19th century to mock any attempt to unify electricity and magnetism, pointing to all the differences between them.

contaminated by matter only later. I will state this in the briefest possible terms without any ifs and buts. While the laws of physics do not favor, to first approximation, matter over antimatter, our actual universe contains matter, not antimatter. This strange fact could be reconciled with the laws of physics through the action of these $12\,X$ gauge bosons required to grand unify. The universe could have started out without an excess of matter over antimatter, or even better, devoid of matter and antimatter. The action of the X bosons could then generate the observed matter content of universe.[51]

A parade of groups

Way back in chapter 1, I mention that Nature's kindness is such that theoretical physicists were taken gently by the hand and guided through what amounts to an undergraduate course on group theory. Allow me to show you the parade of groups in the form of a table. Obviously, I do not expect you to understand the symbols in this table unless you have already learned some group theory.[52] I have to sweep a flood of history out of the way, and instead, I include this table merely to indicate some of the groups that theoretical physicists had to deal with on the road to grand unification. For example, the last line of this table indicates that the grand unified theory based on $SU(5)$ could be extended to one based on $SO(10)$. Why 10? Because[53,54] $10 = 2 \times 5$.

The parade of groups	
spacetime symmetry	$SO(3) \rightarrow SO(3,1)$
global internal symmetry	$SU(2) \rightarrow SU(3)$
local internal symmetry	$U(1) \rightarrow U(1) \otimes SU(2) \rightarrow U(1) \otimes SU(2) \otimes SU(3)$
grand unifying symmetry	$U(1) \otimes SU(2) \otimes SU(3) \rightarrow SU(5) \rightarrow SO(10)$

I can't resist digressing slightly here. Cartoonists often depict physicists wearing lab coats[55] arguing in front of a blackboard covered by complicated mathematical formulas typically taken from "advanced calculus." In fact, for those theorists in the gauge theory game, the blackboard might show $2^2 - 1 = 3$, $3^2 - 1 = 8$, $5^2 - 1 = 24$ for the number of gauge bosons in each of the Yang-Mills theory mentioned in this chapter. And perhaps also $2 + 3 = 5$ and $2 \times 5 = 10$.

Gravity refuses to join in the dance

Finally, finally, we come to gravity. The reader may have heard that physicists have had a lot of difficulty dealing with gravity even though it was the first of the four interactions to be studied. Grand unification has brought the number of fundamental interactions down from four to two! And yet gravity refuses to join in the unification dance.

To start with, the reader should understand that classical Einstein gravity, while continuing to mystify the lay public with its curved spacetime, black holes, gravitational wave, time warp, wormholes, and so on, has been largely understood for decades. To understand all this weird stuff, you don't have to know a thing about quantum physics. The mystery for theoretical physicists has been the challenge of quantizing Einstein gravity. (In this context, I should emphasize that in Hawking's calculation showing that a black hole radiates photons, the electromagnetic field is quantized but the black hole is treated classically.[56] Thus, contrary to a common misconception, Hawking radiation as such does not depend on quantum gravity, although it might eventually shed light on the mystery, as many theorists believe. Hence the tide of interest in quantum black holes.[57])

Let me first clarify a potential confusion. A layreader might wonder why I am allowed to talk about the graviton when all that he or she has heard about is the frustrating inability of physicists to quantize gravity. Well, Max Planck, in his pioneering work on blackbody radiation, which kickstarted quantum physics, effectively theorized that the energy contained in an electromagnetic field was quantized in discrete packets.[58] He was talking about the photon[59] as the quantum manifestation of the electromagnetic field, even though quantum electrodynamics wasn't fully conquered till some five decades later. In the same way, we are entitled to imagine the energy contained in a gravitational field quantized in discrete packets, called naturally and with considerable affection the graviton.

Indeed, even though we do not have a full fledged theory of quantum gravity extending to arbitrarily high energies, we understand quite a few aspects of what we imagine would be the eventual theory. In particular, we could calculate the quantum corrections to gravity,[60] treating it as an effective field theory in the sense discussed in chapter 4.

Gravity is absurdly weak

A major vexation for physicists is that gravity is absurdly weak.

Compared to what? Well, gravity is long ranged, as was first understood by Newton. In apparent solidarity with gravity, electromagnetism is also long ranged. We already mentioned earlier in this chapter that the gravitational attraction and the electric repulsion between two protons both decrease rather gently, like $1/R^2$ as the distance R between the two protons increases, instead of turning off abruptly like the strong and the weak interactions. That the two forces go down in the same way allows us to compare their strengths: take two protons[61] and "measure" the two forces between them. Simply divide one force by the other. The dependence on R cancels out so that this ratio $F_{\text{electric}}/F_{\text{gravitational}}$ is just a number, measured to be about 10^{36}, that is, 1 with 36 zeros after it.

Electromagnetism is stronger than gravity by this humongous factor of 10^{36}. Just crazy! Who would have cooked up such an absurd number? While this mysterious number is crucial for how the universe works (to say nothing about the existence of lumbering giants like us on the scale of the elementary particles), nobody knows where it comes from.

In quantum field theory, that an interaction is long ranged implies that the particle responsible for generating that interaction is massless, as was alluded to earlier. Thus, the graviton, the quantum responsible for gravity, and the photon, the quantum responsible for electromagnetism, are both massless. The puzzle bedeviling physicists is that the coupling of these two massless particles, to mass in one case and to charge in the other, differ so enormously in strength.

Far beyond any conceivable experiments

The gargantuan number 10^{36} implies that the characteristic energy[62] at which gravity "wakes up" is by many orders of magnitude larger than the highest energy attainable at the world's largest accelerators, unfortunately.[63] Our attempt to reveal the true nature of gravity has to be, for the foreseeable future, guided by mathematical elegance rather than by experimental manipulation, and this program, as exemplified by string theory, has not worked out, after a few decades, as well as some had hoped. This also explains the relative (at least in hindsight) ease with which the generation of Feynman and Schwinger dispatched quantum electrodynamics in contrast to the decades[64] of struggle by later generations trying to control quantum gravity—there was a wealth of relevant experiments on quantum electrodynamics.

That gravity is so much weaker than electromagnetism may surprise the unfortunate who just had a hard fall. The reason of course is that every atom in the unfortunate's body is being pulled down by every atom in the entire earth. The enormous number of atoms involved more than compensates the teeny number 10^{-36}.

A fundamental difference is that masses are always positive while charges could be positive or negative. Thus, electromagnetism knows about yin and yang. While the electric force is attractive between yin and yang, it is repulsive between yin and yin, and between yang and yang. In contrast, gravity does not know about yin and yang: everybody attracts everybody else gravitationally.

Common objects are electrically neutral: they contain equal numbers of positive and negative charges. Hence electromagnetism is well hidden in everyday life. Another way of saying this is that the electric force between two everyday objects is the sum total of the attraction between the protons and electrons, the repulsion between the protons, and the repulsion between the electrons. The attraction and repulsion cancel each other out. Whatever force that exists between the two objects is a residual force, left over after this cancellation.[65] It is as if in a financial transaction involving zillions rounded off to the nearest dollar, all we get to see is the rounding error of 23 cents. What electric and magnetic forces we see in everyday life are just the teeny "round off errors."

Higher dimensional spacetime is more than a century old

Interestingly, gravity and electromagnetism as two classical theories were unified already (at least in some sense) way back in 1919 by the German-Polish physicist Theodor Kałuza, who wrote to Einstein to say that he had added a fifth dimension to spacetime. By that time, Minkowski's innovative geometric view of special relativity as a 4-dimensional spacetime bringing together space and time was so clearly true that it had been quickly accepted. Since a fifth-dimension has never been seen, it is assumed to curl up into a tiny circle far smaller than any distance scales that have been probed experimentally. Einstein was quite taken by Kałuza's clever idea.[66] Years later, Einstein confessed, perhaps somewhat ruefully, that dimensions higher than $4 = 3 + 1$ had never occurred to him.

Incidentally, I have told you enough about Einstein gravity and the metric $g_{\mu\nu}$ of curved spacetime in chapter 6 that I could actually convey to you a flavor of how Kałuza's idea works. Generalize $g_{\mu\nu}$, with the indices

μ and ν ranging over 0, 1, 2, 3, to g_{MN}, with the indices M and N ranging over[67] 0, 1, 2, 3, 5. The extra components introduced are evidently[68] $g_{\mu 5}$ and g_{55}. After we compactify (this is the technical term for "curl up") the fifth dimension, physicists in the resulting $3 + 1$ dimensional spacetime (namely, we) would see two extra fields (namely, $g_{\mu 5}$ and g_{55}) cruising about, one carrying a spacetime index μ and hence transforms like a vector under Lorentz transformation, and one not carrying any spacetime indices and hence transforms like a scalar. That vector field gives rise to electromagnetism.[69] That the scalar field has not been identified experimentally is considered by some physicists as a strike against the theory.[70]

The Swedish physicist Oskar Klein rediscovered[71] what is now known as the Kałuza-Klein theory[72] in 1926 and later developed the theory further[73] by generalizing Kałuza's circles to spheres and thus anticipating Yang-Mills theory.

I presume that many readers have heard that string theory is formulated in $10 = 9 + 1$ dimensional spacetime, with 6 of the 10 coordinates curled up. You might say that Kałuza and Klein started this veritable stampede to higher dimensions right around a hundred years ago. Incidentally, if you had caught a whiff of the flavor of these higher dimensional theories as "explained" earlier, you might realize that string theory is beset with a mob of unseen scalar fields, not just one.[74]

The actor and the stage

Physicists have successfully quantized all the fields we know about. In all these cases, the quantized particle is set to move through a preexisting spacetime. A source of deep confusion for theoretical physicists is that in the classical theory of gravity, namely Einstein's theory, which, as I have emphasized, has been well understood for decades with its essential features not in dispute, the particle we know as the graviton and that we are trying to quantize corresponds to fluctuations in spacetime itself. The graviton is thus unlike any other particle, let's say the photon in order to be definite. The photon happily moves through spacetime, whether it is curved or not. The graviton in contrast, is also responsible for controlling the spacetime through which it is moving.

My favorite analogy, which I have used in my other books, is that mysteriously, we find ourselves attending a puzzling play in an avant garde theater, in which one of the actors is somehow also the stage. The graviton itself controls the spacetime it is moving through.

This line of thought inevitably leads to deeper questions about the nature of space and time.[75] For instance, is time an emergent concept? It is fun to talk about this kind of stuff, and many indulge, precisely because there is hardly any constraint on what one may or may not say. Some string theorists have proclaimed that "spacetime is doomed." Certainly, no relevant experiment addressing this issue anytime soon.

Are we required to quantize gravity?

Throughout all this confusion and hair tearing, a small minority has also ventured the opinion that perhaps gravity should not even be quantized. This point of view has been perhaps most forcefully articulated by Freeman Dyson. Over the years, I have slowly gravitated toward this point of view. One simple exercise, which has been carried out by a number of authors, is to contemplate what sort of experiment[¶¶¶] could possibly detect an individual graviton (so that we know for sure that gravity is in fact quantized), and how long it would take. Due to the almost absurd weakness of gravity, as was explained just now, the detection of an individual graviton within a reasonable time duration may require a detector so massive that it is at risk of collapsing and forming a black hole. And if we use a smaller detector not at risk of becoming a black hole, the probability of detecting an individual graviton may be so low that the relevant time scale to see one comes out to be comparable to, or to exceed, the age of the universe. The detailed numbers do not matter, but you get the idea. Gravity is so incredibly feeble that essentially no conceivable experiment could establish definitively that it is quantized or not.

Another issue that greatly bothers me, and many others, is that existing attempts to quantize gravity, such as string theory, blithely assume that the principles of quantum physics as we know them are not modified all the way up to the Planck energy when gravity "wakes up." In other words, regarding gravity's refusal to be quantized, it is always assumed that gravity is "at fault," and not our quantization procedure. I don't see any reason why that should be so. To me, it seems much more likely that quantum physics as we know it would not hold at exceedingly high energies.

[¶¶¶]Analogous to some truly fabulous experiments detecting and manipulating individual photons.

Since this book is after all about the foundational ideas of physics, while these speculative ideas are happily being slung around, I refrain from discussing them further.[76]

Notes

[1] Steven Weinberg, *The First Three Minutes* (Basic Books, 1977).

[2] As was already alluded to in chapter 1.

[3] Quantum tunneling also plays a part in slowing down stellar burning.

[4] Differences abound. Light could propagate in the eternal silence of the space between stars, while sound needs air to get anywhere. See also the film *Wait until Dark* in which Audrey Hepburn played a blind housewife.

[5] A distinguished physicist who served in World War II liked to joke about the three laws relating voltage V, current I, and resistance R he learned in the US Army: $V = IR$, $I = V/R$, and $R = V/I$. He tried in vain to tell the instructing officer that the three laws could be unified into a single law.

[6] Recall the concluding thought in chapter 5.

[7] In one of my textbooks, *FbN*, chapter I.4.

[8] This section is taken in part from *QFT ASAP*.

[9] Too numerous to list here. Indeed, several of my books, including *Unity, Fearful, QFT Nut, and QFT ASAP*, are all devoted to this topic, in part or in whole.

[10] Recall that we already encountered the concept of range in chapter 4.

[11] I prefer the term "gravity wave." See *G*, page 5.

[12] See *G*, pages 26–29 and pages 1–5 respectively. That it took only 129 years from one to the other is quite a tribute to the experimentalists and technologists involved.

[13] Both are discussed in the popular movie *Oppenheimer*. See *Physics Today*, April 2024, for a discussion of the worry over setting the atmosphere on fire.

[14] This section is adapted from *QFT ASAP*.

[15] As a passionate violin player, Einstein would sometimes refer to his pursuit of the unified field theory wryly as "my violon d'Ingres." (If you are not sure about what Einstein meant, see https://wordhistories.net/2018/01/05/violon-dingres-origin/.)

[16] For a description of how weird the weak interaction could be, see *QFT ASAP*, chapter IX.2.

[17] This sentence is anachronistic! Wave functions were not introduced till 1926. Anyway, a detailed and accurate history is not our thing here.

[18] You could readily purchase gauged skirts online: https://www.etsy.com/listing/568603956/gauged-cartridge-pleated-skirt-19th. Gauge also refers to the standard sizes of wine casks. For more details, see *Fearful*, chapter 12, "The revenge of art," and also page 191.

[19] That is, by $e^{i\theta}$, for readers who know about complex numbers.

[20] The technical term is isomorphic.

[21] But in another instance Heaviside was far ahead of his time. In 1888 he realized that a charged particle moving faster than the speed of light in a material medium, such as water, would emit what was much later called Cherenkov radiation, recognized by a Nobel prize in 1958. In case this confuses the reader, Einstein decreed in 1905 that the speed of light c in vacuo sets the speed limit in the universe, but light could, and does, move slower in a medium. See chapter 10.

[22] More precisely, let p and n denote the proton and neutron fields respectively. Somewhere else in spacetime, somebody could refer to $(\cos\theta \, p + \sin\theta \, n)$ as the proton field, and to the orthogonal combination $(-\sin\theta \, p + \cos\theta \, n)$ as the neutron field. (I am simplifying here somewhat.)

[23] This is explained in great detail in my book *Group Nut*. One subtlety that the reader of this book does not have to lose any sleep over is that the $U(1)$ in $SU(2) \otimes U(1)$ is not the same as the original $U(1)$ of electromagnetism, a fact which led some theorists astray. Figuring this out marks the difference between Nobel prize and no Nobel prize. All this occurred before my time. I revered all my elders, including Julian Schwinger, Sid Bludman, Shelly Glashow, Abdus Salam, Steve Weinberg, all of them nice people (or at least nice to me), for their remarkable insight.

[24] In a particularly bizarre kind, when we multiply something by itself, we get nothing. See *QFT ASAP*, page 321, to learn how this kind of multiplication would be relevant for physics.

[25] This may seem like one of those deep mysteries of mathematics, but in fact, almost all the undergraduates in my group theory course could derive this formula without much toil.

[26] See, for example, *QFT ASAP*, pages 178, 196, and 200.

[27] With superscripts $+$, $-$, 0 indicating the electric charges carried by these bosons.

[28] I feel compelled to note that this mechanism was also discovered independently by P. Anderson, by R. Brout and F. Englert, and by C. Hagen, G. Guralnik, and T.W.B. Kibble. For details, see F. Close, *Elusive: How Peter Higgs Solved the Mystery of Mass*.

[29] Some of the others mentioned in the preceding endnote were unfairly shafted by the Swedes. In particular, Gerry Guralnik died of a heart attack shortly after the prize was announced. Robert Brout had already been dead for many years.

[30] Readers wanting more details are referred to *Fearful*, *QFT Nut*, and *QFT ASAP*.

[31] As was already mentioned in chapter 4, this unfortunate choice made by Benjamin Franklin has bedeviled physics and engineering students ever since.

[32] Ultimately in analogy with an electron spinning up or spinning down. See chapter 3.

[33] If you would like to know about the neutrino, a brief account is given in *Fearful*, pages 36–37. The neutrino is the only particle that a major American writer has devoted a poem to.

[34] As we all know, this particular nomenclature in physics is way off: the electron may in fact be the most valuable particle. The original reference was to its mass, negligibly small compared to that of the proton.

[35] Have you ever noticed that dentists tend to be leptoprosopic? I haven't. https://www.ncbi.nlm.nih.gov/pmc/articles/PMC8827515/.

[36] Franklin's choice propagates into the charges assigned to the W bosons.

[37] As a student way back when, I read a book by Feynman titled *The Theory of Fundamental Processes*. Much recommended for those readers who can handle it.

[38] Because emitting a W^+ is equivalent to absorbing a W^- in quantum field theory.

[39] Recall from endnote 32 in chapter 5 that the up quark and the down quark carry the fractional charge $+\frac{2}{3}$ and $-\frac{1}{3}$ respectively. You could check that the process $v + d \rightarrow e^- + u$ conserves charge: $0 + (-\frac{1}{3}) = (-1) + \frac{2}{3}$.

[40] You may wonder why this force is not long ranged. Good question! The answer was arrived at after a long arduous process, but to describe it would take us far beyond this book. A cryptic statement here would have to suffice: Anything with color is supposed to be confined.

[41] This technical term should not be taken literally. It refers to how a given combination of quarks respond to the gauge group $SU(3)$.

[42] For a more leisurely discussion, see chapter V.2 in *QFT ASAP*.

[43] To be fair, people who construct models beyond the standard model are performing the useful tasks of exploring which possibilities are already ruled out. I certainly cannot, and do not dispute this.

[44] Interested in learning gauge theory? If so, I cordially invite you to read *Group Nut*.

[45] Yes, each of these conservation laws follows from a symmetry, as was discussed in chapter 5.

[46] I am talking about a neutron sitting by itself. Some readers might be wondering about the neutrons in an atomic nucleus. Due to the attractive strong force between the neutrons and the protons in the nucleus, the mass of an atomic nucleus is usually less than the masses of all

the neutrons and protons added together, and thus this nucleus would be stable. There are of course also some nuclei that are unstable and prefer to break up into smaller pieces by nuclear fission, as was already mentioned earlier.

[47]This was considered an important puzzle to solve. Indeed, the belief that Roger Dashen had solved it was part of the reason that he was offered a position at the Institute for Advanced Study. Pierre Ramond recalled that in the 1970s Feynman once asked him what was new in physics and so he proceeded to describe some new developments in string theory. Feynman interrupted and said, "I thought that you were going to tell me that something important like the proton neutron mass difference has been solved." For a review of the problem dating from around that time, see A. Zee, *Physics Reports* (1972), pages 127–192.

[48]The curious reader could find a brief explanation in *QFT ASAP*, pages 253–256.

[49]A bullish insistence on total agreement between theory and experiment has on several occasions set physics back. For example, see endnote 7 in chapter 2.

[50]For example, the Chinese had metal, wood, water, fire, and soil. The theory sounds very reasonable to me! For example, from water and soil you could produce wood.

[51]This issue, known as the matter-antimatter asymmetry problem, was a hot topic a few decades ago, but due to the cooling interest in grand unified theory, has been quiescent for some time. For further details, see *QFT ASAP*.

[52]Clearly, there is no way I could convey to you the full subtlety of group theory in a popular book. My textbook, *Group Nut*, weighs in at 614 pages, but the good news is that, to understand the groups mentioned in this table, you only need to read about 200 of these pages. Believe me, it is not difficult. The book begins with the difference between isosceles and equilateral triangles (as was discussed in chapter 5 of this book), and I daresay you have known that for many years!

[53]The 2 is because every complex number contains 2 real numbers! And that was how $SU(5)$ was extended to $SO(10)$ as indicated in the table.

[54]I cannot resist telling you, using the same analogy used in the text earlier, that when we open the box $SO(10)$ comes in, we find a plastic tray with 16 slots to fit the quarks and leptons of one family in. The gauge theory determines the number of slots according to $2^{(\frac{10}{2}-1)} = 2^4 = 16$. See *Group Nut*, chapter VII.1. But we counted only 15 quark and lepton fields before. An extra slot in $SO(10)$? It's exactly where the long lost right handed neutrino field would fit! Coincidence?

[55]Which most experimental physicists do not even wear, let alone airy theorists.

[56]For a more or less accessible textbook treatment, see, for example, *QFT Nut*, pages 290–291, and *GNut*, chapter VII.3.

[57]Which should be distinguished from astrophysical black holes, which are manifestly not quantum, but thoroughly and massively classical.

[58]Actually, it was Einstein who realized later what Planck had done and who told him.

[59]The word "photon" was not coined till decades later, but never mind.

[60]For example, you would expect that the Newtonian energy $E = -GM_1M_2/R$ between two masses separated by a distance R should be modified in the quantum world simply due to the uncertainty principle. See N. E. Bjerrum-Bohr, J. F. Donoghue, and B. R. Holstein, arXiv: hep-th/0211072. See also *GNut*, page 767.

[61]Before elementary particles such as protons and electrons were discovered, any proposed comparison between the strengths of gravity and electromagnetism would have been meaningless. What would you use to do the comparison? Note also that by using two protons to do the comparison, I have biased the result in favor of gravity. Since an electron is about two thousand times less massive than a proton, the ratio of electric to gravitational forces between two electrons would be given by the even larger number $10^{36} \times (2,000)^2 = 4 \times 10^{42}$.

[62]Known as the Planck energy, roughly 10^{19} (10 times the square root of 10^{36}) times the rest energy of the proton. See *FbN*, chapter IV.1. In quantum field theory, the coupling strength varies according to the energy at which the process takes place. At the Planck energy, the interaction between two gravitons becomes significant.

[63] Some physicists have fantasized about what it might take to observe individual gravitons, schemes along the line of building a detector the size of the solar system and watching it for a million years etc. Anyway, the numbers are so absurd that there is no point in working them out or arguing about them.

[64] Contrary to what Heisenberg and Pauli optimistically opined, the brilliant Russian physicist Matvei Bronstein, who was purged and executed at the age of thirty-one in 1938, foresaw this difficulty already, amazingly, in 1935. For more details, see *GNut*, pages 764–765, and *QFT ASAP*, pages 155–156.

[65] This also holds at the atomic level. The hydrogen atom, consisting of a proton and an electron, is electrically neutral. The attraction between two hydrogen atoms to form a hydrogen molecule is a residual effect.

[66] But it did not prevent him from sitting on Kałuza's paper for a year before sending it to the Prussian Academy for publication in 1921. Perhaps Einstein's reluctance was explained in a 1922 letter he wrote to Hermann Weyl, saying that "Kałuza seems to me to have come closest to reality, even though he too fails to provide the singularity free electron." Einstein's dream of seeing the electron emerge as a solution of his field equation has not been realized and seems more remote than ever. The lesson here is not to demand too much of a promising theory.

[67] Recall that, over time, 4 became 0.

[68] In case you are wondering about $g_{5\mu}$, recall that the metric is symmetric under interchange of its two indices and so $g_{5\mu}$ is equal to $g_{\mu5}$.

[69] Namely, the vector potential $A_\mu(x)$ responsible for electromagnetism is identified as $g_{\mu5}(x)$

[70] Although naturally, hundreds of papers have been written about the scalar field.

[71] Independently, the Russian physicist H. Mandel did the same in 1926. For this and some other historical tidbits, see the introduction to *Modern Kaluza-Klein Theory* by T. Appelquist, A. Chodos, and P.G.O. Freund, Addison-Wesley, 1987.

[72] Klein later said that it was Pauli who told him that Kałuza had anticipated his work.

[73] I refer those readers interested in more details to *GNut*, chapter X.1.

[74] Namely the manifestation of the analogs of g_{55}, with 5 extended to 5, 6, 7, 8, 9.

[75] Perhaps the most fascinating of the ideas that grew out of attempts to quantize gravity was a mysterious connection proposed by Juan Maldacena and others, known as AdS/CFT. A theory containing gravity and existing in a curved spacetime known as anti-de-Sitter turns out to be equivalent to a very special kind of quantum field theory existing in flat spacetime with one lower dimension and not containing gravity. Puzzling, eh?

[76] Readers interested in reading more are referred to *GNut*, chapter X.8.

8
THE CREATOR SPEAKS THE LANGUAGE OF MATHEMATICS

> [The universe] ... that great book which ever lies before our eyes ... but we cannot understand it if we do not first learn the language and grasp the symbols, in which it is written. This book is written in the mathematical language, ... without whose help it is impossible to comprehend a single word of it; without which one wanders in vain through a dark labyrinth.[1]

A preamble: how much math do you need for physics?

Students often ask how much math you need to do physics. The only honest answer: "Much much less than you think." (I could have added, sotto voce, "But a bit more than what you know. That goes without saying.") If all you need to score a breakthrough in physics is to know a lot of math, there would be thousands of breakthroughs in physics all the lifelong day, and of course that is not happening. Physics doesn't work that way at all. So my advice: Don't worry about learning math, but soak up as much physics as you humanly can, and then soak up some more. What math you really truly need to advance in physics will come along with the physics.

If you think you already know a lot of physics, learn more. (Goodness gracious, I am repeating myself.) The typical undergraduate physics major may think they know an enormous amount of physics, but the truth is they know rather little. Curiously, many feel that they should master a lot more math. Ironically, the truth may be the opposite. The student already knows enough math to comfortably get by, but needs to go way deeper into physics. To excel in physics, you need to learn more physics.

Bottom line: Don't worry about math, focus on the physics. Learn as much physics as you can, then learn some more. Even more importantly, think deeply about what you already know, or think you know. Quite often, you will realize that you do not really know what you think you know. The necessary math would naturally come along with the physics, and the math that didn't come along is by definition not essential.

While writing these sentences, I ran through in my mind the Nobel certified theoretical physicists whom I know personally. With the singular exception of Roger Penrose,[2] most appear not much versed in mathematics (by that I mean not much more than an outstanding physics graduate student), and if they do, they have the good taste of not showing it, neither in conversation nor in publication.

The remarks above are intended for the young and eager aspiring to contribute.

The unreasonable effectiveness of mathematics in physics

I will actually start this chapter with a thought provoking essay by the Hungarian-American physicist and Nobel laureate Eugene Wigner,* who said that he was puzzled by the fact that mathematics is effective in physics, a puzzling statement that would surely puzzle most layreaders. I then go on to offer my opinion about the relationship between physics and mathematics.

One difficulty in writing this chapter is that what is meant by the word "math" varies enormously in the physics community, let alone among the general public, ranging from garden variety stuff like partial differential equations to the truly esoteric that even many pure mathematicians are not really intimate with. Mostly, I mean by math what most working theoretical physicists mean by math. Here the emphasis is on working, rather than talking.

Born in the minds of mathematicians as pure mental constructs?

Early in my career as a physicist, I encountered by chance an essay by Wigner titled "The Unreasonable Effectiveness of Mathematics in the

---

*Whom we already met in chapter 5.

Natural Sciences." Ever since I first heard about physics, I had taken for granted that mathematics is naturally the language to describe the physical world with. Shouldn't that be obvious? That the gray eminence of the Princeton physics department at that time, and one of the greats of theoretical physics in the 20th century, would puzzle over this fact came to dimmer lights like me as a revelation.[3] I believe that many of my fellow physicists have had a similar experience when and if they encountered Wigner's essay.

To forestall any misunderstanding on the reader's part, I should clarify that, unlike some other university students, physics majors typically do not have time to sit around reading essays musing about their challenging subject. The lay reader thus should not presume that Wigner's essay is in any sense part of the standard physics curriculum. I would venture to guess that, to the contrary, the vast majority of practicing physicists have never read, or even heard of, this essay.

True, much of elementary mathematics, such as arithmetic and geometry, what we might regard as the foundation of mathematics, emerged out of the real world, and so is by construction relevant, and effective, for the real world. But many later developments, such as complex numbers and group theory, were born in the minds of mathematicians, as pure mental constructs, and even, one might say, as just fun and games. Indeed, while inventing group theory in the late 19th century, mathematicians thought that the abstract algebraic structures they constructed[4] could not possibly be relevant to physics, as I have already mentioned in chapter 5. And indeed group theory was not at all essential in classical physics. Neither mathematicians nor physicists could possibly have dreamed that group theory would prove to be indispensable for the development of physics in the 20th century. The story of group theory provides an outstanding example of what Wigner had in mind.

For many years, I chewed over the mystery posed by Wigner. I delighted in telling friends, especially those in the humanities, about how unreasonable it is that mathematics is actually useful in physics, much to their surprise. Decades passed, and one day the African American physicist Ronald Mickens invited me to contribute a chapter to a book[5] he was putting together to commemorate the 50th anniversary of Wigner's essay. I wrote an essay titled "The Unreasonable Effectiveness of Symmetry in Fundamental Physics."[6] Symmetry, expressed in the language of group theory, came to dominate fundamental physics in the 20th century, as we saw in chapters 5 and 7.

What do you mean by the word "mathematics"?: a necessary clarification

Before I plunge on, I have to clarify that the word "mathematics," freighted with emotion for many, means different things to different people, varying enormously according to that person's background. Since you are reading this book, you no doubt have a clearer understanding of what math is and is not than the proverbial guy and gal in the street. By "math" I definitely do not mean what is referred to in the American expression "Let's do the math," uttered soothingly for instance by a realtor to a prospective home-buyer. Rather, math refers to the tightly, logically, beautifully, and elegantly interwoven structures uncovered by the intellects of an infinitesimal sub-set of humans, of which the vast majority of physicists from Maxwell and Einstein onward are not a part.

What math is and is not from the perspective of a theoretical physicist

Sadly, a great deal of confusion reigns among even highly educated human-ities professors about what mathematics is, or more precisely what modern research grade mathematics is all about. There are several levels to this confusion. Bright high school students understand that what the com-mon people call math is actually arithmetic. At a somewhat higher level, they understand that geometry and algebra are all about logical structures, rather than numerical computation.

More advanced students understand that mathematicians are not inter-ested in plugging numbers into the famous, or infamous, high school formula giving the two solutions of a quadratic equation.[7] The interest lies rather in the thousand year struggle to find this formula. The high school kids who memorize this formula thus getting an A+ on an algebra exam do not impress anybody in my circle of theoretical physicists. Math is about understanding the structure of this type of algebraic equation, the quest for the corresponding formula for the cubic and the quartic equation,[8] and the insight that no such formula exists for the quintic equation, as the tragic and romantic[9] genius Évariste Galois tried frantically to set down for posterity the night before he was killed in a duel. Now, that guy really impressed the pecan pie out of everybody I know!

Talking about the tightly interwoven structures of mathematics, I was for years much taken by a story Wigner told in his essay. A layperson sitting next to a physicist, say on a plane, was interested in what kind of math the latter used in her work. So the physicist wrote down on a napkin the so-called Gaussian distribution (to follow the story it is not important to know[10] what that is) and explained to her fellow human that this expression, studied by the great German mathematician Carl Friedrich Gauss, popped up all over physics and mathematics. "It even appears in probability theory," the physicist added, "and hence life insurance companies study this function to predict how long people would live." Meanwhile, the layperson had been examining the napkin. He exclaimed, "Say, isn't that π?" The physicist replied, "Yes, it is." The layperson chuckled, "Yup, I knew all along that you were pulling my leg. What does the circle have to do with life and death?"

Indeed, what does the circle[11] have to do with quantum physics and with quantum field theory? But open a textbook on quantum field theory and you will see π all over the place, not to mention the Gaussian distribution.[12] And even on Julian Schwinger's tombstone, as you may recall from chapter 4.

Math internalized by physicists

In any discussion of the relationship between mathematics and theoretical physics, you have to specify what you mean by mathematics, as I keep repeating. If by math you mean calculus and partial differential equations, then of course, it goes without saying that physics and math are intertwined.

In the history of physics, it has happened on numerous occasions when physicists realized that a previous unfamiliar set of mathematics was needed for progress. Invariably, the relevant math was quickly absorbed by physics. Physicists are not dummies. Indeed, they are by and large a discriminating lot. A body of mathematics deemed inessential, even if introduced into physics, was usually forgotten in due time.

Consider Heisenberg's approach to quantum physics in 1925, based on the mathematics of matrices and operators, a language totally alien to most theoretical physicists at that time. In contrast, Schrödinger's formulation involves a partial differential equation, now known naturally as the Schrödinger equation, something that may sound esoteric to some layreaders but has long been familiar to theoretical physicists since at least the time of the Marquis de Laplace[13] in late 18th century physics. Indeed, after seeing what Schrödinger had done, Heisenberg lamented to a friend that he was not going to get a position. ("The old fogies have not the foggiest idea what a matrix is, and of course they are going to prefer someone who speaks their language.")[14] But now, we all know that, quite to the contrary, he was catapulted to the top rank of theoretical physicists.

This many years later, while the typical layperson or an average high school student would still be blown away by the math that Heisenberg employed, matrices and what-not, as needed for quantum mechanics, that kind of stuff has long been internalized by physics and is by now contained in the toolkit of almost any practicing physicist. Concepts such as hermitean matrices and degenerate eigenvalues are not what I meant by math in this chapter. I am talking about math beyond that, what mathematicians would call math, not what physicists call math.

I also mentioned in chapter 7 that cartoonists often show theoretical physicists wearing lab coats arguing in front of a blackboard completely covered by what looks like esoteric mathematics to the general public, but is typically some gibberish involving freshman and sophomore level math, say an integral of some nasty looking expression involving sine and cosine here and there. I would like to see the lab coats disappear and something like $3 \times \bar{3} = 8 + 1$ (in the lingo, "3 times 3 bar gives 8 plus 1") to go on the blackboard.[15] That would be much more realistic. An outstanding physics student who would breeze through what the general public calls "advanced calculus" with the greatest of ease might have to seriously cogitate to grasp the difference between 3 and $\bar{3}$, and what that has to do with the distinction between quark and antiquark.

What I mean by math, for the purpose of this chapter, is the kind of highly abstract modern mathematics that most physicists do not understand, and in most cases have no hope of understanding, and for many, no desire of understanding. Contrary to what some laypersons think, it is not a matter of being smart or not, but a matter of whether one grew up speaking a certain language and thinking in a particular way. Einstein, for example, might or might not be able to comprehend high level mathematics even if he were willing to devote years to studying that area of abstract thought. (He was also not particularly interested.) One of my mentors, Murph Goldberger, the chair of the Princeton physics department at one time and later in turn the director of the Institute for Advanced Study and the president of Caltech, liked to grumble that some physicists spouted fancy mathematical jargon "merely to frighten small children." I might add "also to impress themselves."

When physics and math grew up together, they were intertwined but later drifted apart

Newton famously had to invent calculus in order to develop his mechanics. By his time, geometry and algebra had been developed for centuries and millennia by Greeks, Indians, Chinese, Persians, Europeans, and others to varying extent. After Newton, physics and mathematics often developed hand in hand. The necessary mathematics for doing a certain piece of physics was invented as needed. Laplace did not read about the Laplacian in some mathematics textbook; he invented it when he needed it. Fourier invented Fourier analysis for the problem of heat transfer he was working on. Examples are legion. Some would say that mathematics and theoretical physics are both constructs of the human mind and hence they necessarily have to be intertwined. I leave this for neuroscientists, perhaps even a philosopher or two, to ponder over. But are they really?

Consider this quote[16] by the eminent 19th century mathematician Charles Hermite: "There exists, if I am not mistaken, an entire world which is the totality of mathematical truths, to which we have access only with our mind, just as a world of physical reality exists, the one like the other independent of ourselves, both of divine creation." Both divine, but perhaps not both accessible by a single mind.

The general public often harbors the mistaken impression that physics and mathematics are almost inseparable. In reality, much of 20th century mathematics had moved farther and farther away from physics. While there are surely those who would loudly disagree with this statement and point to the occasional surges of mutual fertilization between mathematics and physics, such as during some episodes in the development of string theory, but these are rare and far in between. Even in string theory, many emphatically eschew fancy mathematics. Several of the most distinguished practitioners have told me that the truly essential progress, such as the discovery of branes or the realization of a dual structure, did not require much mathematics. To emphasize this point, the late Joe Polchinski, who for many years occupied an office next to mine, once pulled down from his shelves a well known sophomore or junior year level textbook on mathematical methods in physics and told me that to do string theory one should not need much more than what was in that book.

Yes, there are those who proclaim vociferously that some of the concepts underlying fiber bundles and gauge theories are related and arrived at independently by mathematicians and by physicists. Some of these connections may deepen our understanding of gauge theories, but it is fair to say that up till now a knowledge of fiber bundles has not contributed to any of the substantial advances in the physics of gauge theories, such as asymptotic freedom or the relevance of the Higgs mechanism for generating gauge boson masses. The truth is that when Yang and Mills wrote down a gauge theory in 1954 they didn't know a fiber bundle from a woolen scarf. That much is an indisputable fact.

Yes, fiber bundles have no doubt deepened our understanding of gauge theories years and decades later, but to a large extent in an entirely inessential way. It is also fair to say that the probability that the type of physicists who go around mouthing about fiber bundles could have discovered gauge theories in the first place is almost mathematically nil. They are invariably more interested in the math than in the physics.

In contrast, to develop his theory of gravity, Einstein had to learn the differential geometry already invented in the 19th century by Riemann and others. It is also a historical fact that the development of Einstein gravity may not have taken as long as ten years had Einstein been fluent[17] in this branch of mathematics. From time to time, I see a web article referring to Einstein as a mathematician. I chuckle, and I imagine the great physicist does also. I know for sure that this is one article I could pass over. Einstein is revered by physicists for his incomparably deep physical insights, not for his knowledge of mathematics.

Neither too easy nor too hard

So perhaps the next breakthrough in theoretical physics requires mathematics that has already been worked out but which physicists do not know about, or mathematics yet to be invented or discovered. (Famous debate I surely do not want to get into!) Physics may well be stuck in the 21st century and what we need to get unstuck is the mathematics of the 23rd century. In the same way, a brilliant young physicist in the 10th century with some insightful notions about force and motion would be hard put to develop them further without knowing the basic ideas behind calculus, unless he was able to invent them himself like Newton.

But thus far, there is no irrefutable evidence that the progress of physics would have to depend on mathematics beyond the mental capacity of the run-of-the-mill theoretical physicist. Indeed, a deep mystery to me is why the mathematics physicists need to understand the physical world is neither too easy nor too hard, as I've said earlier. It is almost matched fittingly to what a bright physics student, with some effort, could grasp. Certainly, there is no evidence whatsoever that for physics to progress we need the kind of abstruse math that could be mastered only after decades of devoted studies.

The fading of a beautiful friendship and the rise of rigor

Of course if we go way back, a sharp distinction between physicists and mathematicians did not exist. What math you needed for physics you got to invent yourself, as I said earlier. But as physics and mathematics grew and drifted apart, the beautiful friendship faded. String theory fosters hope for a revival of the faded friendship, but as of this moment, the future of string theory itself is not without a shadow of doubt.

The divergence between physics and mathematics, beginning in the late 19th century, has widened ever since. Many physicists blame this dismal tragedy on a stubborn insistence on rigorous proofs that was even denounced by some first rate 19th century mathematicians such as the aforementioned Hermite. He said he wanted to show his students the simple beauty of mathematics while avoiding unnecessary rigor.

Of course, most physicists understand that a minimal level of rigor was necessary to avoid nonsensical results. Even the wildest physicist refrains from dividing by zero, a taboo known to elementary school kids. Another common faux pas involves a forbidden interchange of limits.[18] Nevertheless, most working physicists blithely ignore questions of rigor and simply adopt the slogan "Just do it!" hoping that nothing truly disastrous happens.

Rigor for the sake of rigor, what some physicists call rigor mortis, turns many off.

Perhaps some readers are not fully aware of what mathematical rigor means. Please allow me to tell a story from my freshman year. The first week in the math course I was tested into, we were asked to prove that (see figure 1) any continuous curve going from the upper left corner of a square to the lower right corner and staying inside the square must cross the straight line diagonal joining the lower left corner to the upper right corner. I had no idea how to prove such a glaringly obvious fact, and so I wrote "It is obvious." Well, for that intellectual effort I received a big fat zero. Some readers would understand that what the Princeton pedant expected me to do is to regurgitate scholastically minute and legalistically waterproof definitions of the various notions involved such as "curve" and

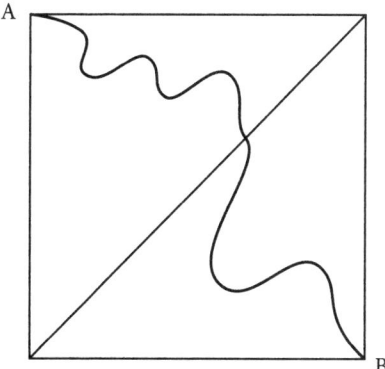

Figure 1. How to prove that a continuous curve going from one corner of the square to the opposite corner must cross the diagonal?

"inside," and to obediently cite the Weierstrass theorem, which he had crammed into our juvenile brains. Most likely, the course was designed by the math department to separate the future physicists from those with the true calling.

A joke here might also help some readers to understand further what mathematical rigor means.[19] An engineer, a physicist, and a mathematician[20] were traveling by train in New Zealand when they saw a black sheep. The engineer exclaimed, "Look, the sheep in New Zealand are black." The physicist objected, saying "You can't claim that! You have seen only one sheep. But wait, if we see another black sheep, or maybe yet another, then we can say almost for sure that all the sheep in New Zealand are black." The mathematician smirked, "All you can say is that, of the sheep in New Zealand we could see from this train, their sides facing the train are black."

The gulf in rigor was vividly enacted for me when I was once invited to lecture at a conference of mathematicians. When I described some work I did with an eminent physicist,[21] a French mathematician sitting in the front row literally flew into a rage, shouting that he couldn't believe that physicists could commit such an outrage. (Actually, as in the joke, it was only a matter of dressing up what I said more carefully in fancier language. I could also understand why mathematicians could get upset over how sloppy physicists and engineers could be in their manipulations.)

In another incident, my Israeli collaborator Feinberg and I obtained a topological result regarding the distribution of eigenvalues of a class of nonhermitean random matrices. (The reader does not have to be familiar with any of these technical words to get the point here.) Some years later, the French mathematician Alice Guionnet and her collaborators[22] proved this result rigorously at a level acceptable to mathematicians. I must confess that her paper is essentially "all Greek to me"; perusing it, all I could recognize is the phrase "Feinberg-Zee Single-Ring Theorem" as she called our result.

I need hardly admit that rigor is essential in mathematics. For example, the announcement in 1994 by the mathematician Andrew Wiles that he had finally proved Fermat's Last Theorem after almost four hundred years of intensive effort by a multitude of mathematicians electrified the entire world. By that time, computers had already checked this fabled theorem for zillions of cases, but that was not a proof of course. The physics community, along with everybody else, stood in awe of that relentless pursuit of an absolute and airtight proof.

But that is mathematics, not physics.

The rise of, excessive some would say, rigor is often associated with the pseudonymous Bourbaki movement in France in the 1930s. This emphasis on absolute rigor then spread to other countries. An English mathematician at the time was said to refer to this trend as the "French disease."[23] Meanwhile, in physics, the pioneers of the quantum world, plunging ahead into the dark unknown, almost to a person are not enamored of mathematical niceties. Indeed, quite the opposite. An exasperated Pauli, who was Austrian himself just like Schrödinger, referred to Schrödinger's work as "Austrian sloppiness" (*osterreichischer Schlamperei*). All I can say is that Heisenberg and Schrödinger were able to birth quantum physics without having the foggiest idea about what exactly a Hilbert space was.[24]

A Fields medalist who read my textbook on quantum field theory[25] once told me that mathematicians indulge, in private, in exactly the kind of heuristic guessing and fooling around that the best theoretical physicists are good at. All those rigorous proofs and fancy language is just a mask to hide the actual thought process, he said. And who am I to doubt a Fields medalist?

Physicists and engineers (and dare I say, perhaps even Fields medal level mathematicians) are hardly alone in looking askance at unnecessary rigor. Allow me to quote two applied mathematicians.[26] "The lack of a rigorous proof should never be an obstacle to constructive analysis. In fact, without [trial and error] there would be very little grist for the mill of rigor. As long as the results ... make sense and can be tested, ad hoc methods can and should be used freely."

Logic is nowhere in sight, let alone rigor

Many epoch making advances in physics involve wild guesses or leaps of faith, as I highlighted in my book *Fly by Night Physics*. For example, Bohr's solution of the hydrogen atom, which kickstarted the quantum understanding of atoms, involves fabricating an equation that we now know is completely wrong.[27] As another example, when students asked me how Planck derived his distribution of black body radiation (never mind what that is if you don't know; it is not relevant to the point being made here), I asked them back, "Derived? From what?" It was just a crazy guess: Planck was committing what he later called "an act of desperation" in his struggle to invent quantum physics (which of course he had no inkling of). In all these heroic incidents, logic was not involved, let alone mathematical rigor.

Sociologically, the big tent of physics accommodates an enormous spectrum of individuals, a continuum ranging from phenomenologists to quasi-mathematicians, as I have alluded to more than once. Indeed, physicists have even been known to win Nobel prizes in chemistry.[28]

Different physicists can tolerate different levels of rigor, no different from the fact that different people can endure different levels of pain. What is rigorous to one physicist might be poetic hallucination to another.

Had that "Schlamperei" Schrödinger been firmly held, before he could publish, to a level of rigor expected by self-appointed mathematical physicists, the world would likely never have heard of the Schrödinger equation: he would surely have been scooped by another even sloppier physicist.

The ocean of mathematics

My understanding, from talking to mathematician friends, is that the ocean of mathematics is now so deep and so vast that what physicists know, with some exaggeration, is but a drop. (And what an individual mathematician knows fills merely a bucket.)

The truly remarkable surprise to me is that until now, even after physicists have understood curved spacetime and black holes, quarks and grand unified theories, there is no compelling evidence at all that a theoretical physicist must master graduate level mathematics in order to understand the physical world.[29] This, we hope, will prove to be another example of Nature's kindness to physicists that I have spoken of repeatedly in this book.

It is quite possible that two or three hundred years from now my statement will be held up for ridicule. Who knows? But for now, I am talking about physics in the first quarter of the 21st century.[†]

The unreasonable effectiveness of undergraduate level mathematics in understanding the physical world

So why is that? Wigner's statement should be amended to the unreasonable effectiveness of undergraduate level mathematics in understanding the physical world.

A deep mystery to me is why the mathematics physicists need to understand the physical world is neither too easy nor too hard for physicists, as I had marveled at earlier in this chapter.

Another mystery is that the laws of physics appear to be matched rather perfectly to human intelligence. To a theoretical physicist like myself, it almost appears as if Nature is taking us by the hand and walking us step by step through what might be considered a not-too-difficult undergraduate math curriculum, at least these days in the early 21st century. (Or perhaps this is an easily explained delusion? That the human mind is capable of only a limited grasp of physical laws and the elementary mathematics that accompanies such a rudimentary understanding?)

One example already discussed: To understand electron spin in the early 20th century physicists had to learn some elementary group theory, just enough to master the group $SU(2)$. This may sound like a lot to people outside physics, but actually amounts to very little[30] in the scheme of things. Physicists at the time, led by the likes of Pauli, had to struggle just a tiny bit. Later, after the discovery of the neutron in 1932, the proposal that there is a symmetry (as was mentioned in chapters 5 and 7) transforming the proton and the neutron into each other once again requires knowledge of the group $SU(2)$. Happy were the physicists! They did not even have to go learn some more group theory.

And a few decades later, when quarks burst on the scene, what was needed turned out to be the group $SU(3)$. Physicists had to simply generalize from transforming two things into each other to transforming three

---

[†]Lest some readers get the erroneous impression, after all my rant and rave about math, that I do not know any math, I would like to state for the record that at present my Erdős number is 3. This means that I had published a paper with A who had published with B who had published with the legendary mathematician Paul Erdős. See the Wikipedia article on the Erdős number and scroll down to see that number for various eminent physicists.

things into each other. From 2 to 3! A jump within the mental capacity of theoretical physicists, especially that of Murray Gell-Mann. (At the time, numerous particle physicists, mainly Europeans and Asians by the way, knew vastly more group theory than[31] MGM, and they were proposing groups far more intricate and advanced than $SU(3)$. For the young readers of this book, herein lies a cautionary tale—who had the last laugh? Still later, the strong interaction that binds quarks together turned out to be based again on $SU(3)$. Again, no need to learn more group theory! How kind is Nature? Too kind.

Another mystery is that physics as we know it has only been around for two or three hundred years, and somehow we can already write down the action governing the four fundamental interactions. (See chapters 7 and 10.) Perhaps I am just sharing the general delusion of the theoretical physics community and kidding myself. Perhaps in the coming years we will be blasted by a tidal wave of mathematics that mathematicians have not even dreamed of. Perhaps, but perhaps not. Wanna bet?

Yes, the Creator definitely speaks the language of mathematics, but yet he, she, they, or it, or whatever, apparently took only the introductory math courses, or quit after a few weeks of the more advanced courses. (Or, perhaps more likely, the Creator knows infinitely more math but declines to show off.) This has happened again and again. For instance, when topology came into particle theory and condensed matter theory, it involves only the baby stuff covered in the first few weeks of a course on topology, such as you cannot lasso a basketball, nor deform an orange continuously into a doughnut—entirely intuitive stuff[32] that you could grasp without the benefit of a math course.

Different areas of physics require different amounts of mathematics

It almost goes without saying that different areas of physics require different amount of mathematics. Some areas need a great deal of mathematics, notably string theory, some other areas not so much, or even almost none at all. (Once again, the measure I use is the amount of mathematics a typical graduate student in theoretical physics, rather than a person in the streets, knows.) But keep in mind what my late colleague Polchinksi said about string theory. At the moment, it is not even certain that string theory as a whole would survive, and if it does, whether those aspects of string theory requiring advanced mathematics would.

One could hardly predict the future, of course, but still, it is difficult to imagine that some of the well established areas of physics, such as atomic and molecular physics, or nuclear physics, would require significant infusions of modern mathematics in order to make further progress.

From the mathematical side, there also appears to be considerable disparity in how well an area could cater to the needs of physics. Calculus turned out to be perfectly suited for the mastery of classical motion, and indeed was invented for this purpose. Differential geometry was almost designed to describe curved spacetime. But other areas of mathematics appear to be ill meshed to what physics needs.

I have, however, a nagging premonition that number theory, which thus far has barely played much of a role in physics, may yet turn out to be important. The ratios of the measured masses of quarks and leptons are for all appearances real numbers with error bars on them, but may well turn out to be large integers or ratios of large integers. Might they be the solutions of Diophantine equations in number theory? (Typically, in contrast to the situation in elementary algebra, there are fewer equations than unknowns, but with the unknowns constrained to be integers.[33] More often than not, the solutions are equal to large integers.)

Mathematics leading physics? Not

One school of thought is that the math needed for the next breakthrough in physics has already been developed unbeknownst to physicists and is just waiting out there to be exploited. As always, nobody knows what that might be until after the fact. Nowadays, some say category theory, but that remains to be seen. As the saying goes, the proof of the pudding is in the eating. We ask not what physics can do for category theory; we only care about what category theory can do for physics. Show physicists one empirically verified prediction that only a knowledge of category theory could provide, and they will elbow each other in a stampede to learn it.

Please do not misinterpret me. I need hardly emphasize that I do not say that advanced abstract mathematics is not a noble and stupendous pursuit, and that it is necessarily not useful for physics. But it also hardly needs to be said that we have to wait to see whether a given piece of mathematics will actually lead to empirically verified advances in physics before we could claim it is useful.

In this connection, it may be sobering to reflect that the two clearly established examples of mathematics once regarded as abstract but now proven to be useful in physics, namely differential geometry and group theory, were introduced into physics over, or close to over, a century ago. (To these one might add notions of matrices and operators now absorbed into quantum physics, and more recently, some elementary concepts in topology.)

The more mathematical among theoretical physicists have always dreamed that pure math would lead us to new physics. But this is still very much just a sweet dream.[34] Historically, there has been no example, at least no major example, of this fabled phenomenon of math leading physics, and here I mean experimentally verified physics of course.

I already mentioned in chapter 4 how James Clerk Maxwell was to synthesize Faraday's experimental research into the theory of electromagnetism. Here is the young Maxwell in his own words: "[I] resolved to read no mathematics on the subject [of electricity] till [I] had first read through Faraday's *Experimental Researches on Electricity*." Some contemporary theoretical physicists enamored of mathematics should take heed! Many in my world dazzle themselves (but not others) with fancy mathematics before they master the underlying physics.[‡]

Maxwell was not led to his equations by some unfathomably abstruse mathematics but by his deep understanding of an outstanding series of experiments and by the profound physical intuition of Michael Faraday, who, as I mentioned in chapter 4, did not know any math due to his lack of an education. Indeed, Maxwell[35] was to consider Faraday's deficiency an advantage, writing: "Thus Faraday, with his penetrating intellect, his devotion to science, and his opportunities for experiments, was debarred from following the course of thought which had led to the achievements of the French philosophers, and was obliged to explain the phenomena to himself by means of a symbolism which he could understand, instead of adopting what had hitherto been the only tongue of the learned." By "symbolism," Maxwell was referring to the notion of field.[36] By "philosophers" Maxwell simply meant "learned men," following the usage of his time.[37] Earlier, Maxwell had said, "the treatises of [the French philosophers] Poisson and Ampère [on electricity] are of so technical a form, that to derive any assistance from them the student must have been thoroughly trained in

---

[‡] A note to students: Maxwell did not say not to read any mathematics. He merely told us the appropriate order of doing things. Physical intuition first, then mathematical analysis.

mathematics, and it is very doubtful if such a training can be begun with advantage in mature years." Indeed, the pace at which sophisticated mathematics has been introduced into string theory and related areas in recent years is such that many physicists "in mature years" would share heartily the sentiments expressed by Maxwell.

The American school of theoretical physics, as exemplified by many but perhaps most vividly by Feynman, has traditionally stressed physical intuition, at the expense of what is sometimes referred to as "fancy shmancy mathematics."[38] I for one was once shouted at, during a lecture I gave at Caltech, by Feynman for "talking math."[39,40]

As was eloquently emphasized by Einstein and as I already mentioned in chapter 4, precisely because Faraday had no way of understanding the mathematics used by the "continental philosophers," he ended up conceiving the all-important notion of field. Many ignorant fools assume the opposite, that if they were to master fiber bundles they would understand electromagnetism. Tell that to somebody like Ed Purcell or Charlie Townes.[41]

This sort of advice, when given to young students intent on going into physics, almost verges on child abuse. An analogy would be to tell someone learning French in order to enjoy more her upcoming trip to France and still struggling to acquire a working vocabulary that it is crucial to first master the subjunctive future conditional tense or whatever. Learn physics first, and then the rigorous mathematical underpinning if you are so inclined.

Einstein was not led to curved spacetime by some mathematical insight, but by Galileo's experiment and by his awe inspiring physical insight into gravity now enshrined in the equivalence principle, as was mentioned in chapter 7. When told by his more mathematical friends of the existence of differential geometry, he learned and used it as a tool, as I mentioned earlier in an endnote. (Again, only the first half of the undergraduate course on differential geometry that I took, not the highly abstruse stuff taught to graduate students in the math department!)

Similarly, Lorentz was not aware of the Lorentz group, but arrived at the Lorentz transformation[42] based on his physical understanding of electromagnetism. He did not propose to generalize $SO(3)$ to $SO(3, 1)$, and never dreamed that $SO(3, 1)$ might be locally isomorphic to $SL(2, C)$, which any number of bozos are capable of mumbling about these days. I could go on and on with an almost countless stream of examples.

How much mathematics does physics need?

A school of thought diametrically opposed to that which opened the preceding section was perhaps most vocally articulated by Feynman, who allegedly declared that if mathematicians did not exist, physics would be delayed by at most a week. This is not an exact quote, but a loose sentiment told to me by my elders when I was starting out. The one week is certainly an exaggeration, characteristic of Feynman if he did in fact say it.[43]

As was mentioned in chapter 6, in the 19th century, the great mathematician Bernhard Riemann, building on the work of Gauss and many others on the geometry of curved surfaces, developed the differential geometry of curved spaces, precisely what Einstein needed to describe curved spacetime as the origin of gravity, as I just said. Incidentally, going from curved space to curved spacetime is a short skip and hop.[44] Even here I have no doubt whatsoever that had Riemann not done his thing physics would not have been set back for too long. The likes of Einstein and his contemporaries are certainly smart enough to eventually figure out what to do, given that they already knew the Pythagoras theorem from their school days and given that Minkowski had already written down the metric of flat spacetime. Who knows how long it would have taken Einstein to develop differential geometry, not all of it, just what he needed. Or, if not Einstein, whose forte was not mathematics, then some other physicist. Several months? A few years? Not long, but likely more than Feynman's week.

As another example, consider Eugene Wigner (yes, that Wigner, puzzling over the effectiveness of mathematics in physics). After his physics doctorate in Germany, he returned to Budapest to take over his father's leather business. But when faced with the prospect of making belts and purses for the rest of his life, he went back in 1926 to Berlin and found a job working for a crystallographer. Then he encountered, while studying crystals, a mathematical problem in group theory and appealed to his childhood friend§ Jancsi von Neumann for help.

> Jancsi considered my group theory problem for about half an hour's time. Then he said, "Jenö, this involves representation theory." Jancsi gave me a reprint of a decisive 1905 article by Frobenius and Schur.

§Note to the reader: Jancsi and Jenö [Johnny and Gene] are what John von Neumann and Eugene Wigner called each other in Hungarian.

... He said, "... it's one of the things on which old Frobenius made his reputation. So it can't be easy." (From Wigner's autobiography[45])

Later, Wigner won the Nobel prize for applying[46] group theory to quantum mechanics. I would like to add that I have taught group representation theory, no longer considered difficult, to undergraduates for close to a decade.[47]

But this story is just background to Feynman's arrogant remark regarding mathematicians. Had "old Frobenius" not existed, how long would it take the likes of Wigner and his physicist friends (such as Heisenberg and Dirac) to develop, not the whole of representation theory, but those pieces needed for quantum physics? Again, more than a week, but not much more.

At the institute for theoretical physics where I work, a postdoctoral fellow who has collaborated with a mathematician pointed out to me one utility in taking some advanced math courses as an undergraduate. "The important thing," she said, "far more than the actual course material, is to learn the mindset of the mathematicians and the jargon they use."

The chasm between physics and mathematics is indeed due in part to the different terminology used. Let me give the reader one rather elementary example. I have mentioned on several occasions two different Lie algebras:[¶] $SO(3)$ (rotations in 3-dimensional space) and $SU(3)$ (transformations of the 3 quarks introduced by Gell-Mann into each other). They are known to mathematicians as B_1 and A_2 respectively.[48] I admit that I am horribly biased, but I submit to the "impartial" reader that the names used in physics make a lot more sense, yes? (To me at least because I grew up with them?!)

Finding all possibilities and classifying them

A type of achievement that mathematicians are capable of, and that theoretical physicists are perhaps not, is the classification of all possibilities. I have always been in awe of the Cartan classification of Lie algebras. Given the definition of a Lie algebra, most theoretical physicists would imagine that there would be an infinite variety of them. The French mathematician Élie Cartan showed that, while there are an infinite number of Lie

---

[¶]As I said in chapter 7, I have thus far intentionally confounded group and algebra to avoid confusing some readers, but here I will make that distinction.

algebras, they could be classified into a relatively small number of families that you can count on the fingers of two hands.

To give you a measure of how difficult this sort of classification is, and how it might be way beyond the capability of a (typical) theoretical physicist, let me mention that the classification of finite groups, worked on by about a hundred mathematicians, was not completed[49] until 2004. Superficially, finite groups look much easier than Lie groups to grasp. A finite group consists of a finite number[50] of elements, while in contrast, a Lie group contains an infinite number of elements. Thus, in a course[51] on group theory for physicists, students are typically told about finite groups first, and then about Lie groups.

Mathematical classification is important to physicists because it tells them, "Hey guys, that's all there is. No need to look around any more."

Some roles of mathematics in physics

Mathematics plays a number of different roles in theoretical physics.

First and foremost, physicists use mathematics as an indispensable tool, indeed, as the essential language, as was already noted.

A second role is to illuminate the physics, usually long after the physics is done and understood. The so-called mathematical physicists, in their most beneficial role, could reveal to physicists deeper structures that they were not aware of. An example of this would be the impact of group theory on quantum physics.

Another example is the formulation of electromagnetism in terms of differential forms. In school, many physicists and engineers have to struggle through Maxwell's equations, but that is partly because the vectorial electric and magnetic fields are much more elegantly and naturally written as a single differential form. But elegant abstract formulations are typically compact precisely because much could be hidden. For sure, Maxwell could not have plumbed the depths of electromagnetism using differential forms. Nor could physicists calculate with them in specific situations.

Yes, I like differential forms and enjoy seeing the elegant way they provide for extracting some of the inner structures of non-abelian gauge theory. No quarrel about that![52] For me, elegance is essential in spite of Pauli's famous dictum to physicists that[53] "Elegance is for tailors."

Another example is the superspace formalism.[54] Sure, one could obtain a supersymmetric Lagrangian by trial and error as was initially done, but it is so much more elegant to invent a superspace with fermionic coordinates for a superfield to live in.

Third, mathematicians and mathematical physicists manage to prove rigorously, often after an arduous struggle, various folk theorems physicists have long believed in and taken for granted. Some of these are surprisingly difficult to prove. An example is the stability of macroscopic matter, defined for this purpose as a collection of negatively charged electrons and positively charged nuclei governed by the electromagnetic interaction and quantum mechanics. Physicists could stand back, admire the tour de force proof, and relax, having been assured that the world has been rigorously proven not to collapse while they sleep.

No go theorems

An important subclass of these rigorously proven theorems are various impossibility theorems, commonly known as "no go theorems," so that physicists would not have to waste time chasing a mathematical impossibility. For example, a particularly important theorem asserts that phase transitions and critical phenomena (an everyday example is the sudden change of liquid water into solid ice) can only occur in systems with infinite number of degrees of freedom.

Thus, the best of these no go theorems prevent physicists from going down dead end alleys. Yet, in many cases, these dead end alleys have been instructive to physicists. Mathematical theorems are necessarily founded on assumptions, of course, and the possible failure, or more commonly, rejection, of any one assumption would provide a dramatic escape from these alleys. My advice to young people is that theorems have to be examined critically. In the history of physics, quite a few no go theorems have been evaded,[55] as I have just alluded to.

Window dressing as "dispensable erudition"

At the other end of the spectrum, mathematical physicists have produced a flood of proofs that, more often than not, most "working physicists" are not terribly interested in, what theoretical physicists refer to as mere window dressing.[56]

A lot of the mathematics touted and bragged about by some in the mathematical physics community tends to be window dressing. An example involves the fiber bundles some people blabber about in connection with gauge theories, as was mentioned earlier. Feynman and his followers (among whom I include myself) strongly dislike this inessential use of mathematics. While Feynman delighted in using vulgar language to describe what he thought of the kind of rigorous mathematical physics he didn't like, let us adopt Einstein's more genteel choice of words: "more or less dispensable erudition." One criterion for Einstein's dispensable erudition is whether someone would miss a chance for immortality in physics, or at least the Nobel prize, because of his or her ignorance of that particular piece of knowledge. Young physicists struggling to make an important contribution to theoretical physics would be well advised to keep that in mind. Of course it is a judgment call!

Fortunately, most working theoretical physicists have the good sense not to pay too much attention to the window dressing. They don't have time to spare. And experimental physicists are even more pressed for time.

One of my favorite biting comments of Wolfgang Pauli is what he said of the much worshipped and absolutely brilliant mathematical physicist John von Neumann, whom you already met in this chapter: "If physics consisted mainly of proofs, von Neumann would have become a good physicist." Einstein had a similarly unflattering opinion about von Neumann, referring to him by the derogatory term "ein Denktier" (literally, "a think animal," or as we would say nowadays, "a thinking machine"). In that spirit, we could say of some physicists that if physics were a branch of mathematics, then so and so might have become an okay physicist.

Talking about window dressers, I am reminded of Niels Bohr's definition of an expert as "someone who starts out knowing something about some things, goes on to know more and more about less and less, and ends up knowing everything about nothing." A window dresser knows a lot of "fancy" stuff irrelevant to actual physics, about something most physicists have long absorbed and regarded as true, on which they have already built an entire edifice. Of course it would be wonderful and dramatic if such a person could show that what physicists assumed to be true were in fact false, consequently causing the edifice to collapse. The physics community would be the first to applaud. But in almost all cases nothing that stupendous happens. The window dresser merely made a display window look "prettier," adding twinkling lights and bird whistles that do nothing for the rest of the edifice while mystifying most working physicists.

My position is perhaps more moderate than that of Feynman and some others. I am willing to admit that there is often a fine line between mathematics that illuminates and mathematics that dresses windows.

"Doing" both physics and mathematics

The real damage is the harmful effect window dressing inflicts on the diffusion of physics among the lay intelligentsia and on the education of the young, the two groups I mentioned in the preface as potential readers of this book. As an author of popular physics books, I receive emails from dedicated and serious readers, both old and young, and it is sad to behold how many are led astray by books written by so-called mathematical physics types. Incredible how many readers believe that they have to master fiber bundles before they could understand gauge theory! Perhaps learning fiber bundles could be a harmless hobby for the retired medical doctors and lawyers and such. But it is literally a scam, at least in my heated opinion, to tell a young student aspiring to be a physicist that he or she has to master a whole truckload of fancy mathematics before tackling contemporary physics.

From time to time, an undergraduate would proudly tell me that he or she is double majoring[57] in physics and in math, or perhaps majoring in physics and minoring in math. My response depends on what that claim means and on the level of sophistication of the student. In many cases, the student has simply taken a couple extra courses on, say, differential equations, or perhaps a course or two on applied math. In those cases, the correct response would be effusive words of praise and encouragement. But rarely, an exceptionally strong student might be tackling quantum field theory and homology theory at the same time. Then, after a few words of praise, I would express my worry about the lack of focus, citing the fact that in modern times, there has been essentially nobody capable of contributing to both physics and mathematics at a high level.

An admittedly inexact analogy might help some innocent students decide what to do. Imagine a person who is able to play American football at say an 80% level of what the players in the National Football League (NFL) are capable of and to play basketball at say an 80% level of what the players in the National Basketball Association (NBA) are capable of. This person's awesome athletic endowment would literally crush the likes of us to bits on the field and on the court. Yet, understandably, professional scouts from the NFL and the NBA would not even bother to look at him.[58]

More often than not, the student also asks for advice about graduate school. Again, my response depends on the level of the graduate school, of which there is a wide spectrum, assuming that we are talking about physics graduate schools. Schools at the lower end would be overjoyed to admit a student who knows more math than what a typical physics major might know. But, I caution the student that I might venture to guess that the top dog grad schools might regard a double major with justifiable skepticism. These top grad schools get to select from candidates who may know nada about homology theory but who would know a lot more about quantum field theory than our cocky double majoring student mentioned above. Whom would you prefer, a student of this type or the one who knows a little bit about both? (There is of course no accounting for taste, de gustibus, et cetera.)

As I said, I am just venturing a guess.[59] (I have certainly not done a detailed study, otherwise I would be enjoying the benefits of an advanced degree in education or sociology.) My eyes would really light up if the applicant had double majored in physics and art history. Indeed, admissions committees at the elite graduate schools would really sit up if the applicant had not taken any course in physics at all, or attended any undergraduate school.[60]

Occasionally, I am invited to give an informal talk to undergraduates in physics, offering advice and such. Whenever I suggest that it would be best to concentrate one's efforts on either physics or mathematics, saying that in modern times it is essentially impossible to excel in both, I run the serious risk of a pedantic professor in the vicinity citing Ed Witten as an example. Well, the fact that Witten is just about the only one over the last fifty or sixty years capable of doing both actually proves my point. The probability that one of the undergrads in my audience has Witten's ability is almost mathematically nil. Furthermore, every string theorist I have recounted this story to has told me that Witten's contributions to mathematics are all based solidly on physical intuition and motivated by physics. Of course, I would be the first to applaud any reader of the book who would be able to contribute significantly to both physics and mathematics at the highest level. Please try.

I believe from my own observations that this sort of false promise that one could do both physics and math has destroyed the careers of more than a few physics students.

A bit of levity

Let me end with a story and two jokes, among many I have heard after spending decades hanging out with theoretical physicists.

During a walk with Heisenberg, Felix Bloch, having just read Weyl's book, proudly quoted Weyl** as saying something like space was simply the algebra of linear operators. "Nonsense," said Heisenberg, "Space is blue and birds fly through it." I trust that the reader realizes by now whether my sympathy lies with Weyl or with Heisenberg.

A mathematician and a physicist were falsely arrested in a totalitarian country for subversion and sentenced to death by firing squad. On the morning of the execution, the warden asked whether the two of them had a last wish.

The mathematician asked for his execution to be postponed to the afternoon. He had long held that all existing proofs of the Pythagoras theorem were not rigorous enough. Miraculously, during the night, he had thought of an absolutely rigorous proof, which he would like to explain to all the prisoners gathered daily for lunch.

The warden said fine, and turned to the physicist, who said that he wished to be shot immediately. No way he could stand listening to a three-hour long rigorous proof of something he had known to be true since elementary school.

In another well-known joke, a physicist flying a hot air balloon drifted through some thick clouds. Upon finally emerging, she had no idea where she was. Luckily, she spied a man on the ground. The physicist lowered her altitude, and cried out to that man, "Kind sir, could you please tell me where I am?" The man lapsed into deep thought, and after a few minutes yelled back, "You are in a balloon." The physicist replied, "Ah, you must be a mathematician. First, you think deeply before coming up with an answer. Second, what you said is absolutely true. Third, what you said is totally worthless."

Notes

[1] Galileo Galilei, *The Assayer* (1623), as translated in *The Metaphysical Foundations of Modern Physical Science* (1925) by Edwin Arthur Burtt.

[2] I may note here without undue hubris that even with his vast knowledge of mathematics, Roger wrote a rather laudatory foreword to my book *Fearful Symmetry*, Princeton Science Library edition, 1999.

[3] Once, while trudging through a snowstorm after dinner to go from the undergraduate dining hall to the physics building, I unfortunately slipped and fell twice. So, when I entered the building I was completely covered with snow. Wigner, about to go outside, asked me, in heavily accented

---

**Namely the befuddled one mentioned in chapter 7.

English, "Is it snowing outside?" For many years afterward, I pondered the unreasonable effectiveness of visual perception in understanding the outside world and the possible profundity behind Wigner's question. Now that we are in another century, perhaps the great physicist was merely being clueless.

[4] Let alone the unfathomable and ever more abstract structures produced by 20th century and 21st century mathematicians.

[5] *Mathematics and Science*, edited by R. E. Mickens, World Scientific, 1990. For the interested reader, the original article by Wigner is reprinted in this book.

[6] See A. Zee, pages 307–323, in the book edited by R. Mickens and cited earlier.

[7] For the benefit of some readers: a quadratic equation of the form $ax^2 + bx + c = 0$ has the solutions $x = (-b \pm \sqrt{b^2 - 4ac})/2a$, but you certainly do not need to remember this to grasp the point I am making.

[8] A cubic equation has the form $ax^3 + bx^2 + cx + d = 0$, a quartic equation, $ax^4 + bx^3 + cx^2 + dx + e = 0$, a quintic equation $ax^5 + bx^4 + cx^3 + dx^2 + ex + f = 0$, so on and so forth. Here a, b, c, d, e, f denote generic numerical constants. Only the quadratic equation is commonly taught in American schools.

[9] Some authors claim that the duel was provoked by rivalry over a certain Mademoiselle du Motel, but more likely it was because Galois was a Republican.

[10] If you must know, it is just $e^{-x^2}/\sqrt{\pi}$. Here are two famous quotes (taken from a forthcoming book by Paul Nahin) about the integral of that function, known as the Gaussian integral: "It is natural to an engineer to ask a mathematician for a finite formula for [the indefinite Gaussian integral].... If we fail to satisfy him it is not because of our stupidity, but because the world does not happen to have been made that way."—G. H. Hardy (1877–1947), a great English mathematician, in a frank admission that not all integrals are "doable." "A mathematician is someone to whom [the definite Gaussian integral] equals $\sqrt{\pi}$ is as obvious as twice two is four is to you."—a quote (perhaps apocryphal) said to be the words of William Thomson (1824–1907), aka Lord Kelvin, to a class of almost surely perplexed students. Actually, Kelvin's definition of a mathematician is rather passé. That the definite Gaussian integral equals $\sqrt{\pi}$ is by now obvious to those of us who learned Poisson's amazing trick in school.

[11] In chapter II.2 of *QFT ASAP* I explain that quantum physics is sort of like classical physics wrapped in a circle.

[12] Indeed, it defines what is known as free field theory. See, for example, *QFT Nut*.

[13] The title was awarded in 1817 during the Restoration, which showed the political acumen of someone born in 1749.

[14] Heisenberg's feelings have been repeated by generation after generation of young theoretical physicists.

[15] As was mentioned in chapter 7, this provides the mathematical basis for one of Gell-Mann's great contributions: the 3 represents the quark, the $\bar{3}$ the antiquark.

[16] Cited by Gaston Darboux, *Eloges académiques et discours*, Hermann, Paris, 1912, page 142.

[17] In particular, his ignorance of the Bianchi identity confused him greatly. Einstein was fortunate (physics also) to have a friend, Marcel Grossmann, who was able to teach him differential geometry. I had mentioned in chapter 6 the untimely death of Hermann Minkowski.

[18] My favorite example is the incomprehension of some high energy theorists of how electric resistance could possibly arise if the underlying electromagnetic interaction is defiantly time reversal invariant. This apparent puzzle is resolved by the different orders of taking limits in nonrelativistic physics. See, for example, footnote 6 on page 366 of *QFT Nut*.

[19] I mentioned this particular joke in my textbook *Group Theory in a Nutshell for Physicists*.

[20] Recall also the joke in chapter 4 making fun of the mathematician's inability to distinguish the real world from the ideal world.

[21] And published in one of the oldest academic journal in France.

[22] A. Guionnet, M. Krishnapur, and O. Zeitouni, arXiv:0909.2214[math]; see also her lecture at the 16th International Congress on Mathematical Physics, held in Prague, Czech Republic, 2009.

[23] A term that has long meant syphilis in England.

[24] The mathematician and mathematical physicist John von Neumann, whom we already met in chapter 5 and will meet again in this chapter, told quantum physicists some years later that what they were dealing with was known to mathematicians as a Hilbert space. In my opinion, incorporating this mathematical language in an introductory course on quantum mechanics, as is now often done, merely adds to the difficulty some students have in learning the physics without illuminating much. (According to one story, when the mathematician David Hilbert, whom we met in chapter 6, visited the Institute for Advanced Study, he told people he was desperate to talk to von Neumann to find out what a Hilbert space was.)

[25] In which I consistently thumb my nose at mathematical rigor.

[26] H. P. Greenspan and D. J. Benney, *Calculus: An Introduction to Applied Mathematics*, 1973, page 122.

[27] Otto Stern and Max von Laue, two physicists who both later won Nobels, took an oath to quit physics if Bohr's "nonsense" turned out to be correct, as I already mentioned in chapter 3. See also *FbN*, page 24.

[28] When my esteemed and beloved former colleague, the late Walter Kohn, for whom the building I worked in is named, won the Nobel prize in chemistry, the physics faculty jokingly voted to expel him from the physics department.

[29] We are not talking about publishing highly mathematical papers on string theory.

[30] I could teach it to a bright undergraduate who knows linear algebra in an hour or so.

[31] MGM told me that he was struggling with $SU(3)$, roughly at the level of a student halfway through my undergraduate group theory course. I didn't believe him.

[32] Such as that covered in *QFT Nut*, chapters V.6 and V.7.

[33] I am thankful to Greg Huber for many discussions of Diophantine equations.

[34] In contrast, as I understand it, Ed Witten was awarded a Fields medal for showing what physics could do for mathematics.

[35] In contrast to Faraday's up-from-rags background, Maxwell, as the scion of a distinguished family, received the best education that his era could provide, and was thereby able to achieve the grand mathematical synthesis of electromagnetism.

[36] Actually called "lines of force" by Faraday.

[37] In the sense that Anglo American physicists are PhDs, doctors of philosophy.

[38] I will refrain from exploring the historical and sociological origins of this emphasis, which has been, at once, the strength and weakness of American physics. Generally speaking, European physicists receive a much more vigorous and extensive training in contemporary mathematics than their American counterparts. The French philosophers, now referred to as the French physicists, are still regarded by many Americans as way too mathematical. (I have actually published with several of them.) Of course, what is considered fancy by one generation is often thought basic by the next. The mathematics used by Poisson et al. now looks like child's play, and is familiar to any undergraduate student of physics.

[39] Lest some readers think that Feynman is somehow "not good at math," let it be noted that he won the 1939 Putnam Exam, an annual national mathematical competition held in the United States.

[40] Just to forestall some readers thinking that I am against the excessive use of abstract mathematics in physics because I don't know any math, I like to note here that I had in collaboration with a pure mathematician published in a mathematical journal a paper using notions like "Freudenthal suspension."

[41] If you don't know who these two giants were, please Google. Suffice it to say that they are among the physicists from the second half of the 20th century I most admire.

[42] Several 19th century physicists, including W. Voigt and J. Larmor, came close to or derived the Lorentz transformation before Lorentz. See *GNut*, chapter 6, page 169. What counts is not the math, but the physics implied by the transformation, and this was appreciated by nobody else but the one and only Albert Einstein.

[43] I am not completely sure about that, but if not in those exact words, certainly the gist of it.

[44] *G*, pages 143–150.

[45] From "The Recollection of Eugene P. Wigner (as told to Andrew Szanton)," cited earlier.

[46] Among Wigner's many contributions to theoretical physics: he introduced his sister Margit to the notoriously eccentric and shy Paul Dirac, who might otherwise have ended up like Newton. For a biography of Dirac, see G. Farmelo. To counterbalance somewhat the standard image of Dirac as "the strangest man," as Farmelo put it, see also the article by F.-C. Chen (*Mod. Phys. Letters A* 2020002), which contains a posthumous speech given by Mrs. Margit Dirac.

[47] And "old Frobenius" has become one of my favorite mathematicians.

[48] A translation dictionary is given on page 385 of *GNut* for the four infinite families of Lie algebras.

[49] One highlight was the discovery of the Monster group with about 8×10^{53} elements.

[50] For example, 8 elements.

[51] And in textbooks, such as my *Group Nut*.

[52] See, for example, B. Zumino, Y.-S. Wu, and A. Zee, *Nucl. Phys.* v. 239 (1984), pages 477–507. I would never have been able to derive the more intricate results in this paper using the clunky notation used ordinarily in physics.

[53] I believe that he might have said this in jest, or just to be ironical and contrary. One of his best known works is his elegant explanation of the accidental symmetry in the hydrogen atom using the Lie algebra $SO(4)$.

[54] For a simple introduction, see, for example, *QFT Nut*, chapter VIII.4.

[55] A famous case is the Coleman-Mandula theorem, which was circumvented by the invention of supersymmetry, which subsequently spawned an entire industry in theoretical physics and quite possibly the most expensive wild goose chase in human history.

[56] To use an American idiom referring to department stores and meaning to make things look better than they are. Incidentally, and somewhat off topic, the French expression for window shopping means "licking the windows."

[57] My apology to those readers not subject to the American educational system; the meaning, however, should be self evident.

[58] If some readers feel that I am verging on the emotional here, it is because in graduate school at Harvard I helplessly watched some highly gifted friends spiraling into oblivion. Most of them had an almost hopeless struggle finding suitable employment.

[59] But based on my experience serving on the admissions committee for graduate studies in physics at Princeton decades ago.

[60] Examples of the former include E. Witten, of the latter F. Dyson and R. Phillips, at present or formerly professors at the Institute for Advanced Study and at Caltech respectively.

9
ENTROPY AND THERMAL AGITATION: ALL ABOUT SHARING

A mysterious concept

Which concept in physics has mystified physicists for the longest time? Ask a physicist or a historian of physics. You may be surprised that many would say temperature.

After Newton, physicists knew how planets move and cannonballs fly, but hot and cold? Mysterious! Take for example the infamous "caloric fluid." Another one of the many concepts stuffed long ago into the dustbins of history! So if you have never heard of caloric fluid, nothing lost. For ages past many sages thought that heat was an invisible fluid contained inside hot bodies, but physicists have long left that concept for others to chew over.

The ultimate understanding of temperature had to wait for the introduction of the profound concept of entropy by the German physicist Rudolf Clausius in 1865. He initially used the German word *Verwandlungsinhalt*, meaning "transformation content," which mercifully did not catch on, and later coined the word entropy from the Greek word for turning* and for transformation. But yet the concept of temperature did not get fully clarified till the molecular basis of matter, and even more importantly, quantum physics, was fully understood. (I will get to this later in this chapter.) This difficult concept has continued to baffle generations of physics students; it certainly confused me; more than relativity and quantum mechanics, I might add without much exaggeration.[1]

The final conquest of thermal physics was achieved only through the monumental work of the Austrian physicist Ludwig Boltzmann, one of the

---

*Recall the Tropic of Cancer from which we got the word tropical? Another interesting word with the same root is zoetrope, an ancestor of the movie, based on the notion that turning gives rise to animation and life (zoe as in zoo and zoology.)

all time greats (a GOAT as Americans would say) of theoretical physics. Crucially, he battled his contemporaries, furiously championing the notion that the world was atomic and molecular rather than continuous. He suffered drastic mood swings[†] throughout his stormy life, ending in suicide by hanging in 1906.

Some historical background and a disclaimer

Two intertwining threads run through physics and technology in the 19th century. During an industrial revolution powered by the steam engine, the need to understand the physics of heat and expanding gases took on an additional urgency. At the same time, the molecular basis of matter was becoming ever more plausible. Using this molecular picture, physicists could readily understand the macroscopic properties of a gas, such as the pressure it exerts as a function of temperature.

I mean to give here a cursory and idiosyncratic introduction to this rather difficult subject,[2] with various subtleties and niceties suppressed and swept into the closet. The curious reader is referred to the many textbooks on the subject, ranging from the willfully obscure to the preternaturally lucid, with not a few mathematically rigorous treatises.

Consider a gas in a container, perhaps outfitted with a piston that the gas could push against to do work. Macroscopically, the gas is characterized by several physical quantities, including its total energy E, the volume V of the container, the pressure P it exerts on the walls of the container, its temperature T, and the amount of heat it contains. But what does the word "temperature," distinguishing hot and cold and thus known to every child, actually mean? And similarly the word "heat"?

Physicists eventually began to suspect that the gas actually consists of zillions[‡] of tiny molecules zipping around hither and thither in largely empty space. This microscopic picture of a gas, now taught to schoolchildren, was met in the 19th century with fierce opposition by those who did not accept the outrageous claim that apparently continuous matter was

---

[†]Boltzmann himself joked that his personality was due to his being born the night between Mardi Gras and Ash Wednesday in 1844. Nowadays, we would say that he suffered from bipolar disorder. One story was that when he was invited to lecture at Berkeley toward the end of the 19th century, he was outraged to discover that Berkeley was a dry town and that to have the beer and wine indispensable to his Mitteleuropa lifestyle he had to sneak off to neighboring Oakland.

[‡]Meaning some humongous numbers of order 10^{24}.

actually empty space populated by zillions of teeny, and likely fictitious, entities.

Clausius and others[3] eventually postulated that the age-old concepts of heat and temperature were ultimately derived (as I will explain later) from a mysterious entropy, denoted by S in reputable textbooks. In their views the molecules are incessantly colliding with each other and bouncing off the walls of the container. As any billiards player knows, in a collision of two balls, each carrying more or less the same momentum, one could emerge moving much faster, the other much slower. Or vice versa. The microscopic state of the gas, characterized by zillions upon zillions of numbers, describing the position and momentum of each molecule, is changing instant by instant, as collisions between molecules occur.

Evidently, it is not necessary, nor is it possible, to keep track of this mind-bogglingly large number of variables. The macroscopic variables that experimentalists could measure and listed above, such as energy E and pressure P, are given by the sums of microscopic variables averaged over time. In this sense, this branch of physics is known as statistical mechanics or statistical physics. That physicists do not need to, nor care to, track each and every molecule is similar to the sociologists' indifference to the fate of each and every individual.

The total energy of the gas is clearly not going to be shared equally among the molecules: some are moving fast, while others are moving slowly, just as in a human society, some are rich and some are poor. Physicists speak of a curve, known as a Maxwell-Boltzmann distribution, specifying what fraction of the molecules have energy a little bit above the average, way above average, a bit below average, way below average, and so on.[4] It turns out, as you might have guessed, the temperature T of the gas is determined by the average energy of the molecules: sort of like the per capita wealth of a country.

Thus, each macroscopic state, characterized by a handful of quantities, corresponds to an almost inconceivably large[5] number of microscopic states.[§] Call this number W. Boltzmann figured out that Clausius's

---

[§]Here's another one of my pet rants. A subject in physics, known as classical statistical mechanics, makes sense to me only as a well defined approximation to quantum statistical mechanics as Planck's constant tends to zero. The word statistics implies counting, but how can we count if energy does not take on discrete values? I know of course that some physicists could hem and haw and dance around this issue.

mysterious entropy S is simply the logarithm[¶] of W, and famously, had his fundamental formula $S = \log W$ carved on his tombstone in Vienna.

Now, imagine what happens if we pump energy into the gas, commonly referred to as "heating the gas." Statistically, the energy of each molecule will increase. There will be fewer slow pokes among the molecules, but nevertheless, still always a few. The distribution shifts toward higher energy, again sort of like what happens when a country becomes wealthier.

Disorder and the second law of thermodynamics

Entropy turns out to measure the disorder of the gas. This relationship is remarkably difficult, not only for students of physics but also for physicists in general, to wrap their heads around, as I had mentioned. Intuitively and heuristically, however, it seems plausible that with increasing temperature, that the disorder will increase.

Indeed, let us imagine setting up the gas in an exceedingly ordered microscopic state in which we have somehow arranged for all the molecules to have the same energy, but with half of them moving to the east and the other half moving to the west, such that they stream past each other without colliding. This corresponds to a state with exceptionally low entropy. Never mind how we could even imagine arranging such an unusual state, but clearly, the slightest disturbance would grow explosively and destroy the order. For instance, suppose one molecule in a billion, instead of moving strictly east or west, moves defiantly at a slight angle. Almost immediately, there will be collisions, and there will be collisions. Each collision send molecules zipping off in different directions. In a short time, the molecules will be moving about in every possible direction, but yet in absolute obeisance to the Maxwell-Boltzmann distribution. In this simple example, we could see heuristically the working of the world famous second law of thermodynamics: Disorder, and thus entropy, will always increase until it attains its maximum value. This has been bandied about so much in the popular media that it has more or less entered into the general culture.

Incidentally, thermal physics, merely by carrying different names, has confused students and layreaders of popular physics books. Generally speaking, physicists refer to the understanding in terms of atoms and

---

[¶]Which I will explain later for the benefit of those readers who have forgotten that time is the logarithm of money.

molecules post Boltzmann as statistical mechanics (mechanics because it has to do with the motion of particles), and the earlier stuff as thermodynamics. I prefer the name thermal physics, and favor statistical physics over statistical mechanics since nowadays we are no longer restricted to the motion of particles. Much of the research on condensed matter physics involves fields, and cosmologists study the behavior of quantum fields at high temperatures, such as in the universe shortly after birth.[6]

Classical versus quantum gases

An important point. Since classical physics deals with continuous quantities and since entropy involves counting, the concept of entropy is, strictly speaking, not well defined in classical physics. Only the difference in entropy between two macroscopic states can be defined.

Statistical physics makes sense only with the coming of quantum physics,[7] in which we talk about discrete states that we could count with our fingers.[8]

Into the quantum era! The statistical physics of quantum gases was worked out by a number of luminaries, including Planck, Fermi, Dirac, Bose, and Einstein, with profound consequences for physics and perhaps even humanity, giving rise to such technological marvels as superfluidity and superconductivity. The conquest of statistical physics in the quantum domain also turned out to depend on various fundamental concepts, such as the identity and indistinguishability of quantum particles.[9]

Information and entropy

Now we switch gears somewhat, and describe an apparently unrelated development coming from left field. Toward the end of World War II, while working at the Bell Telephone Laboratories, the applied mathematician Claude Shannon[10,11] wrote a classified memo on the mathematical theory of cryptography, and thus formulated a quantitative measure of information.[12] While information theory is clearly relevant to computer science, and in particular to the possible development of a quantum computer, it remained remote from the mainstream consciousness of physics for a long time until recent decades, when it was recognized to be intimately related to entropy. (Indeed, physicists now think that information and entropy are equivalent concepts.) One main line of research on quantum gravity leads to whether or not black holes preserve the information content of the infalling matter.

To understand temperature, we have to understand entropy, and to understand entropy, we have to understand information theory.

To explain the concept of information in physics, I will tell you a fable.

A fable about a princess bride

In many ancient societies, India and China for example, upper class women were typically secluded until their parents arranged to marry them off. Imagine a young woman eager to gain some information about the man soon to be her husband. She asks her brother to go on a reconnaissance mission and report back whether this man chosen by her parents is hand-some or ugly. (As a first pass, let us adopt the crudest approach and assume that the available men in the village are equally likely to be handsome or ugly.) In this way she obtains valuable information. The question is how much.

It is to Shannon's enormous credit that he posed this question and gave a quantitative answer.

Next, suppose our curious young woman wants to know more and asks her informant to say whether the man is handsome, ordinary looking, or ugly. (Again, assume that the three categories are equally populated.) Clearly she now has more information than before. Think of the brother's report as the answer to a multiple-choice question with w choices, again assuming that each of the w choices is equally likely. The amount of infor-mation, call it I, should increase as w increases. Write the information as a yet unknown function of w: $I = f(w)$. The only thing we know about this function $f(w)$ at this point is that it should increase as w increases. In our example, when w goes from 2 to 3, the amount of information I should also increase: $f(3)$ is larger than $f(2)$.

You could readily think of many more examples. The more boxes (for instance, marital status: single, married, separated, divorced, widowed, etc. etc.) you have to check on a government form, the more information you are giving them. Also, clearly, if there is only choice $w = 1$, then there is no choice at all: so if $f(1) = 0$, no information is transmitted.

At this stage, we (or more accurately, Shannon) do not know what the function $f(w)$ is.

Now our young woman becomes even more discriminating. She asks her spy to also find out whether her future husband is intelligent or dumb (once again, assuming the two possibilities are equally likely). Evidently, the report she subsequently receives will contain even more information.

Since the question about looks has three possible answers and the question about smarts has two possible answers, her spy has to answer a multiple-choice question with $6 = 3 \times 2$ possible choices: handsome and intelligent, handsome and dumb, ordinary looking and intelligent, ordinary looking and dumb, ugly and intelligent, ugly and dumb.**

Our young woman would not like ugly and dumb, but could perhaps tolerate ugly and intelligent. Whatever her preference, she will receive more information, equal to the sum of what she would have gotten if she had asked about looks and intelligence separately. We conclude that the unknown function $f(w)$ must be such that $I = f(6) = f(3) + f(2)$, again assuming that the choices are equally likely for simplicity.

Remarkably, Shannon realized that these required properties about information sufficed to determine the unknown function f. Abstracting away from our fable, we require that

$$f(wv) = f(w) + f(v)$$

for any two integers w and v. The function we want turns multiplication into addition!

The logarithm searched for and found

A bell may ring for some readers. Toward the late middle ages, all kinds of people dealing with numbers, merchants, accountants, builders, and so on would have wanted to turn the more laborious process of multiplying two numbers together into the simpler process of adding two numbers together. This drive eventually led to the invention of the logarithm[13] in the 17th century by John Napier and others. The function that Shannon was looking for is none other than the logarithm: $f(w) = \log w$.

Indeed, all kinds of properties of the logarithm can be deduced immediately from the defining relation

$$\log(wv) = \log w + \log v \quad \text{(defining relation for the logarithm)}$$

For example, set $v = 1$. Then $\log(w \times 1) = \log w = \log w + \log 1$. It follows that $\log 1 = 0$, as was noted earlier: with only one choice, no information is transmitted. Similarly, set $v = 2$ and conclude that $\log(2w) = \log w + \log 2$.

**Here we are assuming for the sake of simplicity that looks and brains are not correlated, which may well be contrary to fact.

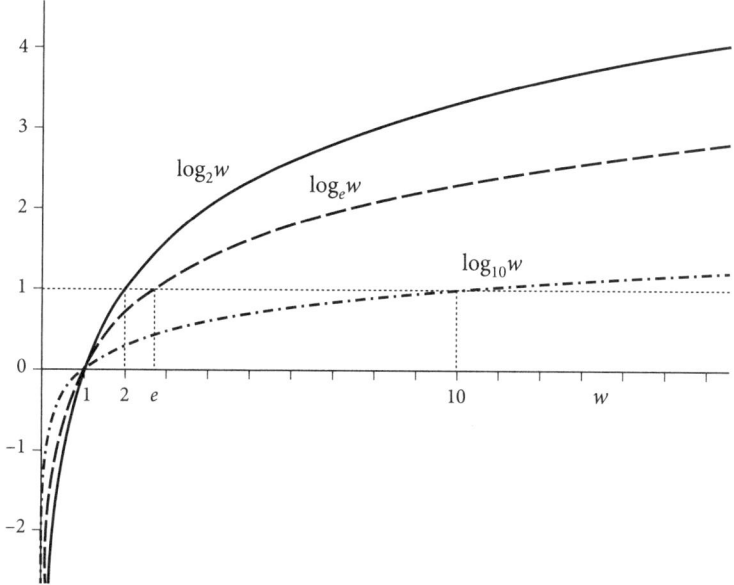

Figure 1. The logarithm is plotted for base 2 and base 10, and for its natural base $e \simeq 2.718$, with the base indicated. A standard notation is to indicate the base b by a subscript on log: thus, by definition $\log_b b = 1$ as you can see in this figure for $b = 2$, e, 10 respectively. Note that regardless of base, log 1 vanishes. The logarithm function log w is notorious for its slow increase with w. By log, I will always mean $\log_e$.

Since $2w$ is larger than w, $\log(2w)$ is larger than $\log w$ and hence $\log 2$ must be positive.

The difference between $\log(2w)$ and $\log w$ is simply $\log 2$ regardless of how large or small w might be. We see that the logarithm increases rather slowly. For instance, as you increase w from 1,000,000 to 2,000,000, the information $I = \log w$ goes up by only a measly $\log 2$. (In physics, a logarithmic increase is almost synonymous with barely an increase.) See figure 1.

I should mention a minor (and unimportant for our purposes) technical detail here: the base b of the logarithm is defined to be that number such that $\log b = 1$. Then according to the defining relation for the logarithm $\log b^2 = \log bb = \log b + \log b = 2 \log b = 2$. Similarly, $\log b^3 = 3$, $\log b^4 = 4$, and so on.

High school students are usually taught logarithm to base 10, because we have ten fingers. It makes a lot of sense since, in base 10, the logarithm of $10 = 10^1$ is equal to 1, the logarithm of $100 = 10^2$ is equal to 2, the logarithm of $1,000 = 10^3$ is equal to 3, and so on. Consequently, engineers normally use logarithm to base 10.

In our computerized age, most readers probably know what a bit is and that computer scientists use base 2 and count with two fingers,[14] known as 0 and 1.

But we would hardly expect the laws of the universe to refer to the number of fingers we have. When the day comes for us to discuss physics with extraterrestrials, who knows how many fingers they have? Perhaps neither 10 nor 2, but none. That's why physicists use what is called natural logarithm, for which the base is the universal mathematical constant $e = 2.71828\cdots$ (known as Euler's number) instead of the number 10. Note that e is not even an integer. Once again, let me emphasize that this potentially confusing business about base is not that important for our purpose in this chapter.

When I talk about logarithm, I will always mean natural logarithm.[††] Thus, for us, $\log e = 1$ by definition of e, $\log e^2 = 2$, $\log e^3 = 3$, et cetera. From this, we could interpolate to define a function e^x, known as the exponential function, for x not necessarily an integer, by the relation

$$\log e^x = x$$

This indicates that the logarithmic function is the inverse of the exponential function: if you evaluate the exponential of x, and then take the logarithm of that, we get back x. Similarly,

$$e^{\log y} = y$$

(Indeed, if you apply the logarithm to the left hand side of this equation, you would obtain $\log e^{\log y} = \log y$ according to the preceding equation, but $\log y$ is just the logarithm of the right hand side. So it works.) In other words, the exponential and the logarithm undo each other.

I assume that the reader is familiar with everyday terms such as "exponential growth," and bank advertisements touting that through the "magic of compound interest" money grows exponentially with time: $M \sim e^t$. This leads to the famous saying that time is the logarithm of money: by taking the logarithm of the preceding equation, we obtain[15] $t = \log e^t \sim \log M$. As is also well known, the time it takes a capitalist to grow his net worth

---

[††]Often denoted by ln, but I prefer to use log.

from \$10 billion to \$20 billion is the same as the time for him to go from \$1 billion to \$2 billion.

Be that as it may, the function that Shannon was looking for is just a natural logarithm!

Information as a function of probability

Back to our story about the princess bride. For ease of exposition, I assumed that in her local population handsome, ordinary looking, and ugly men were equally numerous. Thus, her spy has three choices to fill out, $w = 3$, and the information conveyed amounts to $\log w = \log 3$. But in real life, these three types of men do not occur with equal likelihood, and instead we have to deal with probabilities. To illustrate, and for the sake of definiteness, suppose that our bride-to-be regards only 10% of men in her community as handsome and thinks 30% of them ugly, leaving 60% branded as ordinary looking. In other words, the probability that her husband is handsome, ordinary looking, or ugly equals $p = 0.1$, 0.6, and 0.3 respectively, We have to generalize the formula $I = \log w$ for information.

It is not difficult to guess an appropriate generalization. In our simplified example, the probability that the husband is handsome, ordinary looking, or ugly is just $p = 1/3$ equally. (More generally, $p = 1/w$.) It is natural then to generalize the formula for information from $\log w$ to $\log 1/p$, by replacing w by $1/p$.

Thus, we (or rather Shannon) deduced that the information conveyed by knowing that an event with probability of p of occurring has actually occurred equals

$$I(p) = \log 1/p$$

Note that we have not proven this but have simply argued that this reduces to the previous case when p happens to equal 1 divided by an integer. (For the interested readers who can manipulate logarithms, I outline a derivation in an appendix.)

Let us check that this makes sense. If the bride's spy tells her that her husband is handsome, the information conveyed would be $\log 1/0.1 = \log 10$, while if the bride's spy tells her that her husband is ordinary looking, the information conveyed would be $\log 1/0.6 = \log 10/6 = \log 1.666 \cdots \simeq 1.67$. Since our earlier discussion indicates that the logarithmic function $\log x$ increases steadily, albeit slowly, as x increases, $\log 10$ is larger than $\log 1.67$. Not only is our bride happy to learn that her husband is

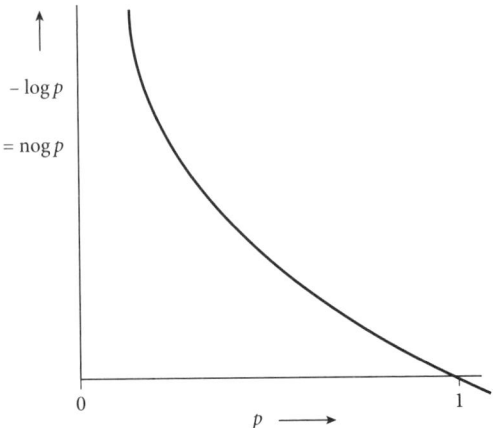

Figure 2. Plot of information $I = \text{nog } p = -\log p$ as a function of the probability p. The amount of information decreases sharply as p increases from 0 to 1. Note that this figure is just a portion of figure 1 inverted.

handsome rather than merely ordinary looking, she has also gained more information.

The more unlikely an event, the more information is conveyed when we are told that that event has in fact occurred. Makes total sense!

An alternative way of writing information

In this section, I mention an alternative way of writing $I(p)$ by using a property of the logarithm that follows immediately from the defining relation:

$$\log p + \log 1/p = \log p \times (1/p) = \log 1 = 0$$

Thus, we deduce that $\log 1/p = -\log p$. From this, we also see that for p a positive number less than 1, $\log p$ is negative and hence $(-\log p)$ is a positive number, as we already know by construction.

Summary: If p denotes the probability of receiving a message, then the amount of information contained in that message is $I = -\log p$. See figure 2.

This pesky minus sign is simply due to the fact that the logarithm of a number less than 1 is negative while the amount of information is positive by definition. If you do not like the minus sign, you could write equivalently $I = \log 1/p$.

Another possibility is define a new function ("negative log") nog by $\text{nog}\, p = -\log p$ so that $\text{nog}\, p$ is a positive function for p in its natural range from 0 to 1, decreasing rapidly from infinity as p increases, reaching 0 at $p = 1$.

Let us point out several features of this plot. For $p = 1$ (corresponding to $w = 1$, only one choice), we have $I = 0$, no information. This makes sense as was already pointed out repeatedly. Suppose I tell you that tomorrow the sun will rise in the east. Given that the probability of this occurring is extremely close to 1 (if not exactly equal to 1), you have gained zero information from me. Consider the other extreme. Suppose I tell you that California will sink into the ocean tomorrow (never mind how I knew and whether I could be trusted). The probability of this actually occurring is close to 0, and as you can see from the figure, the information you have gained is enormous, close to infinity: $I = -\log 0 = \text{nog}\, 0 = +\infty$! If somehow I secretly knew, I would have given you a huge amount of information. Sell some of your tech stocks immediately!

These two extreme examples serve to emphasize that Shannon's formula doesn't tell us how reliable the source of that information is, nor how valuable is the information, just the amount of information transmitted. Information engineers only care about how many bits could be transmitted per second, but not about the artistic quality of the movie being streamed.

The more unlikely an event, the more information if we find out that event actually occurred

I am getting pretty close to the concept of entropy, but yet another example from "real life" might be helpful. Late in the fall at the institute for theoretical physics where I work, an administrative assistant would have sorted all the letters of recommendation for the hundreds of applicants for the handful of postdoctoral positions and placed them in boxes. (I am talking about the bad old days; nowadays all that is put into an enormous computer file to which the professors are given the password.) What is the total amount of information contained in those boxes?

Suppose we label the different levels of recommendation by i and write the probability that a given letter would be of type i as p_i. For instance, a particular i could correspond to "His or her engine runs." I am not making this up; an eminent physicist at an elite east coast university likes to summarize his letter of recommendation with an automotive image. "His engine runs" means "This applicant received a PhD at our institution and he is basically competent." Many of the letters are of this type, saying essentially that this applicant is run-of-the-mill ordinary, so the corresponding p_i is rather high, say 0.9, fairly close to 1. On the other hand, the information nog $0.9 = \log 1/0.9 = \log 1.1111 \cdots = 0.10536 \cdots$ contained in each one of these letters is rather small (these numbers are of course just illustrative, but I give them to show that Shannon has made the rather vague notion of information quantitative). But there are so many such letters! Hence the total amount of information (suitably normalized) contained in letters of this type is equal to $p_i(-\log p_i) = p_i \text{ nog } p_i$. We multiply the amount of information $(-\log p_i) = \text{nog } p_i$ contained in each such letter by the probability p_i that any specific letter would contain that amount of information.

In contrast, consider an extremely rare type of letter saying that this applicant is comparable to a truly great like Dirac.[16] Everybody is of course overjoyed to see a letter like this. The information contained in this letter is enormous because the corresponding probability is so low, close to 0. Again I am not making this up. The letter[17] Oppenheimer wrote for Feynman said something like "He is the new Dirac except that he is human."

The more unlikely the letter, the more information it contains. Indeed, by now you know well that the smaller p_i, the larger the amount of information $(-\log p_i)$. But if we want to calculate the total amount of information, we have to take into account the rarity of such letters, and weigh the information by p_i.

So, add up all the numbers $(-p_i \log p_i)$ to obtain the total amount of information contained in those overflowing boxes

$$I(\{p_1, p_2, p_3, \cdots\}) = -(p_1 \log p_1 + p_2 \log p_2 + p_3 \log p_3 + \cdots)$$
$$= \sum_i (-p_i \log p_i) = \sum_i p_i \text{ nog } p_i$$

The symbol Σ_i consisting of the capital Greek letter sigma denotes summation, as usual in mathematics. I have also written out the sum explicitly to avoid any misunderstanding. It instructs us to take the information $(-\log p_i)$ contained in letters of type i, weigh it by the probability p_i that such a letter would be in those boxes, and sum the product over all possible types. If you think that this is quite a mouthful when stated in English, then that is exactly why the laws of physics are written throughout the universe in the language of mathematics (as was discussed in chapter 8).

Connection to physics

Now back to physics and that proverbial container of gas. By the end of the 19th century, physicists, including Boltzmann and the American physicist Josiah Willard Gibbs,[18] had worked out an expression for the entropy of a gas. As I mentioned earlier, the energies of the zillions of molecules follow the Maxwell-Boltzmann distribution. To make contact with the preceding discussion of information, we have to do some minor adjustment. For a classical gas, we speak of the fraction $f(E)$ of molecules with energies between E and $E + dE$ with dE infinitesimally small. This plays the same role as the probability p_i in the preceding discussion. In other words, the discrete variable i corresponds to the continuous variable E, and p to f. Thus, the sum over the discrete variable i is replaced by an integral over the continuous variable E.

Incidentally, all this talk about replacing sum by integral etc. is not of conceptual importance for our purposes here. In fact, we now know that the world is quantum in character, and so we have sums, not integrals, in any case. It is just that for expository purposes, I need to place the reader back in the late 19th century. Anyway, the formula for entropy late 19th century physicists obtained has the schematic form

$$S \sim - \int dE \, \rho(E) \, f(E) \log f(E)$$

(Here $\rho(E)$ is a factor known technically as the density of states, but need not concern us here.)

Well, well, it should not have escaped the reader's notice that with the correspondence of sum and integral etc., this is essentially what Shannon had derived as information I:

$$I = \sum_i (-p_i \log p_i)$$

Looking at these two expressions, you could see that what Shannon called information is intimately related to what Clausius called entropy.[‡‡] Indeed,

---

[‡‡] According to a story said to be an urban legend, Shannon went to ask von Neumann (whom we already met in chapters 5 and 8) what he should call the quantity he had been studying. The latter suggested entropy "because that is what it is," but "even more importantly, since no one knows what entropy really is, then in a debate you will always have the advantage." But Shannon ignored von Neumann's advice.

upon more careful examination, physicists came to realize that the two are in fact the same!

In particular, how does this relate back to that proverbial container of gas? We feel that with all these zillions of molecules bouncing around every which way, any microscopic arrangement of the molecules is as probable as any other. Thus, each microstate would have a probability of $1/W$, where W denotes an outrageously huge number equal to the total number of microstates. Shannon's formula for information then gives[§§] $I = W(1/W)\log W = \log W$, which is exactly the formula Boltzmann had carved on his tombstone in Vienna.

Order versus disorder

Let me give a heuristic handwaving motivation for identifying entropy with information. People associate entropy with disorder: the more of a mess, the more entropy. The second law of thermodynamics stating that entropy always increases is then the statement that disorder always increases in the absence of external interference.

First, a tiny bit of math. Consider the function $\sigma(p) \equiv -p \log p = p \operatorname{nog} p$ as p ranges from 0 to 1. Consider the two extremes. Since $\log 1 = 0$, we have $\sigma(1) = 0$. The other extreme is a little trickier. As $p \to 0$, $-\log p \to \operatorname{nog} 0 = +\infty$, as indicated earlier. But in $\sigma(p) = -p \log p$, the logarithm $(-\log p) = \log 1/p = \operatorname{nog} p$ is multiplied by p, which is tending toward 0. As usual, in these school problems of what is $0 \times \infty$ equal to, the issue is who gets to 0 or ∞ faster. Does p vanishes faster or does $\operatorname{nog} p$ blow up faster?

As explained earlier, and as shown in figure 1, as $p \to 0$ the logarithmic function $\log 1/p$ increases so slowly that a logarithmic growth is often regarded as no growth at all. (Recall that the logarithm of 20 billion is equal to the logarithm of 10 billion plus $\log 2$.) The factor of p going to 0 wins easily,[19] so that $\sigma(0) = 0$. The behavior of $\sigma(p)$ as a function of p is shown in figure 3.

Now consider the value attained by $S = \sum_i \sigma(p_i)$ if all but one of the p_i's vanish. The one that does not vanish has to equal 1 since the sum of all the p's equals 1. According to the preceding discussion, and denoting the number of possible i's by N, we obtain $S = (N - 1) \times 0 + \sigma(1) = 0$.

[§§] All the p_i's are equal to $1/W$ so that each term in the sum is the same and equals $(1/W)\log W$. The sum has W terms, and hence we obtain the expression cited in the text.

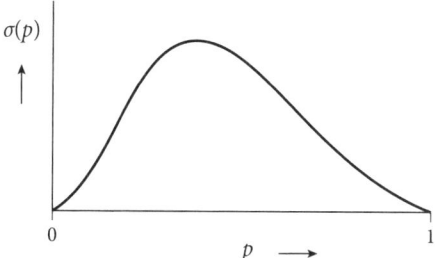

Figure 3. Behavior of the function $\sigma(p) \equiv -p \log p = p \, \text{nog} \, p$ as p varies from 0 to 1. Notice that it rises from 0 at $p = 0$ and then drops back down to 0 at $p = 1$, reaching a maximum at $p = 1/e$. To obtain the total entropy S for a given set of $\{p_i\}$, we simply add up $\sigma(p_i)$ as explained in the text.

The system is perfectly ordered. Indeed, the entropy vanishes. But there is also no information. (In our fable, imagine that the princess bride already knows for sure that her husband-to-be is rich because her father has financial dealings with his father.)

What about the opposite case? Suppose all the choices are equally likely, so that for all i, $p_i = 1/N$. Think of N as a substantial number, say 9 or 10 to be definite. The system is completely disordered, and so the entropy should be high. Then $S = \sum_i \frac{1}{N} \log N = N \times \frac{1}{N} \log N = \log N$; the second equality here follows since the sum consists of N terms of equal value. Indeed, the entropy S is substantial if N is substantial, certainly larger than 0.

One source of potential confusion is that when all the messages are equally likely it might seem that the princess bride would have received no information. In fact, it is the opposite because, of the N possible messages, she has received a specific one. She has gained a lot of information, namely an amount of information equal to $\log N$, and the larger N is, the more information. Remember, the information contained in a given message means the information actually acquired by a recipient after she reads that message.

Disorder	High entropy	Lots of information
Order	Low entropy	Little information

Biological life, entropy, and information

In 1944, Schrödinger, one of the fathers of quantum physics, published a relatively thin but hugely influential book with the audacious title[20] *What Is Life?*, in which he stressed the intimate connection of biological life with entropy and information. In essence, life could be defined as a system that is intricately designed to dump entropy into its environment, and to extract negative entropy[21] from the environment. This understanding had spurred Watson and Crick, among others, to search for how genetic information is coded in biological systems, leading to DNA. More recently, the role of information in biology has been increasingly emphasized.[22]

Statistical or thermal physics

In classical physics, energy is a continuous quantity, but in quantum physics, as we learned in chapter 3, energy is quantized and can take on only certain discrete values, different for different systems. So, consider a quantum system whose energy can only take on the values $E_0, E_1, E_2, \cdots, E_i, \cdots$. Imagine a collection of N such systems (known as an ensemble[23] in physics), with total energy NE to be shared among these systems. The number of systems N is taken to be enormous, say something like 10^{24}, or more precisely, a number approaching infinity, thus accounting partly for the origin of the word "statistical" as I said earlier.

What is the number of systems, call it N_i, with energy E_i? In other words, what is the probability $p_i = N_i/N$ that a given system has the energy E_i, given that the total energy equals NE? (Note that we care about p_i, but not really about N_i and N; they were introduced for ease of exposition.)

As a student, whenever I encountered words such as "system," my brain would blank out for a moment: this class of "generic" words always struck me as too vague. So, I would like to be a bit more specific here. The standard example given to students is a container of gas, as I had already mentioned earlier. In the present context, each of the zillion molecules zinging around corresponds to a quantum system. The ensemble is the gas. According to quantum mechanics, the energy of each molecule could take on a set of discrete values E_i determined by the size of the box, the molecule's mass, and so on.

The Boltzmann distribution

So, how do you determine p_i? If you haven't taken a course on statistical mechanics, you wouldn't have the foggiest idea. Neither did physicists before the likes of Maxwell, Gibbs, and Boltzmann came along.

Well, our earlier discussion suggests that as the gas molecules collide and exchange energy, the entropy or disorder increases until it cannot increase any more. The total energy is conserved even while individual molecules gain or lose energy.

So, let us hear Boltzmann thundering: Maximize the entropy S subject to the two constraints

$$\sum_i p_i = 1 \quad \text{and} \quad \sum_i p_i E_i = E$$

The first constraint just follows from the commonsense everyday Las Vegas definition of probability: probabilities must add up to 1. The second statement just says that the average energy in the ensemble is fixed to be E. Or, if you prefer, conservation of matter (in the form of gas molecules) and conservation of energy respectively.

Well, engineers, economists, businesspeople, and many others can perform this kind of optimization with constraints: it is an easy problem in multivariate calculus. So can most undergraduates in a course on statistical mechanics. The answer is remarkably simple.[24] The probabilities p_i are determined to be

$$p_i = \frac{1}{Z} \, e^{-\frac{E_i}{T}} \quad \text{(Boltzmann distribution)}$$

This result for p_i is known quite appropriately as the Boltzmann distribution.[¶¶] The exponential function $e^{-\frac{E_i}{T}}$ has appeared, together with two constants T and Z we haven't seen before. We should perhaps not be surprised by the appearance of the exponential, given that it is the inverse of the natural logarithm, as I remarked earlier, and that the definition of entropy involves the logarithm. Also, it is not surprising that two constants T and Z pop up since we have two constraints, no more and no less, to fix them.

What temperature means

The Boltzmann distribution tells us that systems with E_i much larger than T are exponentially unlikely. For instance, the probability that a given system has energy equal to T is given by $\frac{1}{Z} \, e^{-1} = \frac{1}{Z}\frac{1}{e} \simeq \frac{1}{2.718Z}$ while the probability that a given system has energy equal to say $10T$ is given by $\frac{1}{Z} \, e^{-10} = \frac{1}{Z}(\frac{1}{e})^{10} \simeq (\frac{1}{2.718})^{10} \frac{1}{Z}$. The relative probability equals $e^{-10}/e^{-1} = 1/e^9 \simeq (\frac{1}{2.718})^9 \simeq 1/8{,}095$. Having an energy equals to $10T$

[¶¶]It generalizes the earlier Maxwell-Boltzmann distribution for a gas. Simply substitute for E_i the kinetic energy of an individual molecule.

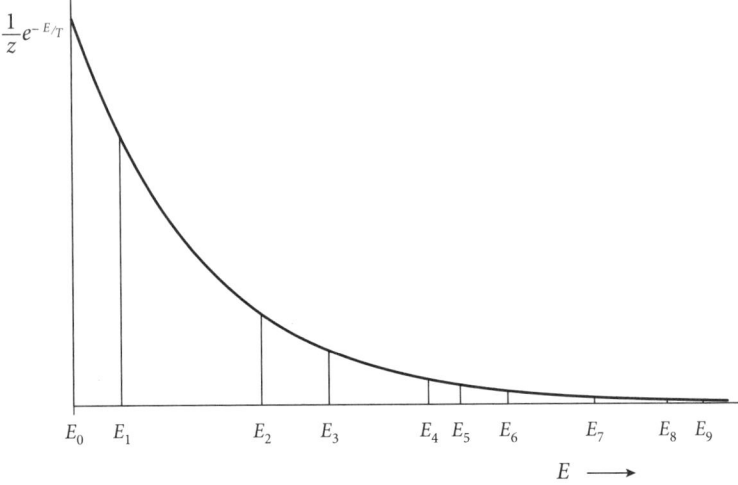

Figure 4. The Boltzmann distribution for a quantum system with energy levels E_1, E_2, E_3, $\cdots$: the curved line represents the function $Z^{-1}e^{-\frac{E}{T}}$, while the vertical bars have height $Z^{-1}e^{-\frac{E_i}{T}}$ respectively. As the temperature T goes down, the states with higher energies are populated less and less likely.

is almost ten thousand times more unlikely than having an energy of T. In other words, the exponential function $e^{-\frac{E_i}{T}}$ decreases rapidly as the ratio $\frac{E_i}{T}$ increases, as indicated in figure 4.

The quantity T governs the distribution of energy, and how unlikely extreme energies are. It is related, to the average energy E, but in a way that depends on the spectrum E_i's.

To summarize, the typical molecule has an energy roughly equal to T, while molecules with energy far exceeding T are exponentially rare. The Boltzmann distribution thus generalizes the Maxwell distribution, being applicable to gases as well as to other thermal systems.

More trumpet blasts please. A triumph of 19th century physics, the mystery of what temperature meant was solved! Using the Boltzmann distribution, physicists can show that this quantity T has all the properties expected of the concept of temperature. For instance, schoolchildren learned that temperature was first quantified by filling a tube with mercury (more commonly with alcohol nowadays since mercury is highly toxic), which expands or contracts according to the ambient temperature. The expansion of matter is contained in the Boltzmann distribution. As T increases, the average energy of the molecules increases, and as the molecules move about more energetically, they cause the liquid to expand.

Bottom line: this quantity T in the Boltzmann distribution corresponds to our inborn notion of hot and cold, to what humans from time immemorial have called temperature.

Temperature clearly should be measured in units of energy. Unfortunately, as a historical accident so to speak, it is commonly measured in degrees, which are merely markings on a tube of alcohol known as a thermometer. A conversion factor, known as Boltzmann's constant and denoted by k, is needed to convert degrees to temperature. Certainly, k is not a fundamental constant such as c and $\hbar$. In the text, I have effectively set k equal to 1. In some areas of physics, unfortunately, people persist in using degrees, even though temperature expressed in energy units is much more informative. For example, in early cosmology, to say what the temperature of the universe at a certain epoch was in energy units (rather than so many zillion degrees) immediately tells us whether atomic physics, nuclear physics, or particle physics is relevant.

Bridge between the microscopic and the macroscopic worlds

Boltzmann's epoch marking work provides the bridge between the microscopic world of atoms and molecules and the macroscopic world of gases, solids, and liquids. Given the Boltzmann distribution for p_i, we can now calculate the entropy by simply plugging in the formula $S = \sum_i (-p_i \log p_i)$. This means that we can now calculate this mysterious entropy which had puzzled physicists for centuries. Indeed, since we know the probability distribution, we could calculate almost any macroscopic quantity experimentalists can measure, such as the total energy E, in terms of the microscopic quantities E_i, namely the energy levels specified by quantum mechanics. Indeed, an introductory high school level course on probability and statistics would tell us that $E = \sum_i p_i E_i$, as was stated earlier. Thus, the entropy S of a container of gas is a function of the energy E, the volume of the container V (through the E_i's as noted earlier), and other physical quantities (such as the mass of the gas molecule). Alternatively, we can also think of E as function of S, V, and so on. This is all a mouthful for saying something quite simple, namely that we want to focus on the function $S(E)$ or equivalently the function $E(S)$, keeping all other quantities, such as V, fixed. (Hence we don't even write the dependence of these functions on them.)

The key question turns out to be asking how S and E vary, with all other quantities held fixed. For instance, as we pump energy into the system, how does the entropy change? More precisely, if E changes infinitesimally by the amount dE, how much is dS, the infinitesimal change in S? The ratio $\frac{dS}{dE}$ then measures how the change of one quantity affects the other. Alternatively, we could consider the ratio $\frac{dE}{dS}$. I regret having to be so wordy, but I want to be precise about what we are talking about.***

You might have noticed that the word temperature and the letter T do not appear in the two preceding paragraphs. Hold your breath! We are about to discover what temperature means.

Well, Boltzmann told us what S and E were, and Newton and Leibniz taught us how to differentiate. Following these greats, we evaluate the derivative $\frac{dE}{dS}$, and lo and behold, that elusive concept of temperature T pops out. Indeed, the age-old field of thermodynamics (going back to alchemy) now emerges, in particular its central formula

$$T = \frac{dE}{dS}$$

relating temperature, energy, and entropy.[25] Temperature measures how the energy changes when the entropy changes. Equivalently, we could write this in terms of the two infinitesimal changes:

$$dE = TdS \quad \text{or} \quad dS = \frac{dE}{T}$$

The confusion, which befuddled physicists for so long, and which continues to befuddle physics students, is because the two concepts, energy E and temperature T, are so easily, indeed, so naturally, mixed up. In everyday life, we know that as we pump energy into a gas, it becomes hotter. We also intuit that a hotter gas contains more energy. Thus, we could be excused for thinking that somehow E and T are the same thing. In fact, the formula $T = \frac{dE}{dS}$ shows that while the two are intimately related, they are definitely not the same.[26]

The astute reader might have noticed that I have avoided using the word[27] "heat": that opens up yet another can of worms!

***Readers conversant with calculus would know that this is known as a partial derivative, meaning that we take the derivative of one quantity against another holding everything else fixed, and should be written as $\frac{\partial E}{\partial S}$. But in order not to confuse other readers, I will continue to write the ordinary derivative $\frac{dE}{dS}$.

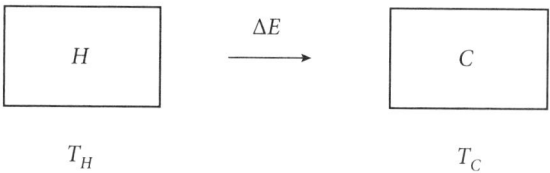

Figure 5. Energy flows from one system, labeled H and at temperature T_H, to the other, labeled C and at temperature T_C. The system H loses entropy while the system C gains entropy. For $T_H > T_C$, there is a net increase of entropy in the universe.

Two thermal systems sharing energy

For immediate use here, keep in mind the relation $dS = dE/T$: it tells us how much the entropy of a system must increase if you wish to increase the energy of a system. For a given dE, you divide by T to obtain dS. In particular, the higher the temperature T of the system, the smaller the increase in entropy dS for a given increase in energy dE. In other words, the hotter the system, the less the cost in entropy increase for it to absorb a given amount of energy.

Consider two containers of gas transferring energy from one to the other. See figure 5. They are said to be in thermal contact with each other. (In the figure, for the sake of clarity, the two containers are not actually shown to be in contact.)

Denote by $\Delta E > 0$ an infinitesimal and positive amount of energy that flows from system H to system C. (You will soon see what the subscripts H and C refer to.) According to the central formula of thermodynamics we just discussed, system H lost an infinitesimal amount of entropy ΔS given by $\frac{\Delta E}{T_H}$, while system C gained an infinitesimal amount of entropy given by $\frac{\Delta E}{T_C}$. Note that we have used the conservation of energy here! Balancing its account book, the universe discovered its entropy has changed by

$$\Delta S_{\text{net}} = \left(\frac{1}{T_C} - \frac{1}{T_H} \right) \Delta E$$

If T_C is less than T_H, then $\frac{1}{T_C}$ is larger than $\frac{1}{T_H}$ and $(\frac{1}{T_C} - \frac{1}{T_H})$ is positive. Thus, ΔS_{net} is positive. The universe has gained a net amount of entropy!

Indeed, H and C refer to hot and cold respectively. The second law of thermodynamics, that the entropy of the universe always increases, is demonstrated by the empirical observation that energy always flows from a hot object to a cold object, never the other way around.

I should emphasize that a rigorous proof of the second law is difficult and controversial, no doubt contributing in part to Boltzmann's frustration and suicide. Here we have to be content with at best a heuristic motivation.

A hot body is eager to share its energy

There are various ways of looking at the central formula of thermodynamics. One way is to ask how much entropy dS a body with temperature T has to give up when it loses an amount of energy dE. Since[28] $dE = TdS$, a hot body could give you lots of energy with a relative small cost in entropy. Thus, a summary of thermodynamics:

> Temperature measures the eagerness of a hot body to share its energy with you.

The second law is equivalent to the statement that energy tends to flow from hot body to cold body, as noted earlier. It is exceedingly unlikely for the bottom half of your mug of coffee to become frozen while the top half starts boiling. Energy conservation alone does not forbid the bottom half from steadily transferring energy to the top and getting colder and colder in the process. We need the second law of ever increasing entropy to forbid this peculiar event from occurring.

By now, you realize that the concept of temperature, unlike the concept of energy, is statistical in character, or, as some would say, derived. It is simply a measure of how energy changes when entropy changes. One way to underline this is to ask how hot is an individual electron. You could ask how much energy an individual electron carries, but you are not allowed to inquire about its temperature.

Temperature is the property of a collection of electrons, not of an individual electron.

Hawking radiation from black holes

With this understanding of entropy, you are now in a position to understand Hawking radiation from black holes. In classical Einstein gravity, a black hole is an extreme warp in spacetime such that objects that fall in could never get out again. John Wheeler was among the first to realize that there is something fishy about this textbook story. He understood that if a black hole happened to come by while he was drinking tea, he could violate the second law of thermodynamics by throwing his tea into the black hole. I will let Wheeler tell, in his trademark style, the story[29] of how he conveyed this disturbing thought to his graduate student Jacob Beckenstein:

> One afternoon in 1970, . . . I told [Beckenstein] of the concern I always feel when a hot cup of tea exchanges heat energy with a cold cup of tea. By allowing that transfer of heat . . . I increase (the universe's) microscopic disorder, its information loss, its entropy. The consequences of my crime, Jacob, echo down to the end of time, I noted. But if a black hole swims by, and I drop the teacups into it, I conceal from all the world the evidence of my crime. How remarkable! Beckenstein, a man of deep integrity, takes the lawfulness of creation as a matter of the utmost seriousness. Several months later he came back with a remarkable idea. You don't destroy entropy when you drop those teacups into the black hole. The black hole already has entropy and you only increase it!

At the time of this fateful encounter, not only between the cup of coffee and the black hole, but also between John Wheeler and Jacob Beckenstein, nobody suspected black holes to have entropy. The reader has to understand that the Schwarzschild black hole is simply a solution to some partial differential equations in Einstein gravity, as was explained in chapter 6. On the other hand, the discussion in this chapter shows that a system possesses entropy if it has an internal structure defined by a characteristic sequence of energy levels.

The key point is that after Beckenstein asserted that a black hole had a finite entropy, then according to the formula $T = \frac{dE}{dS}$ you and I "found," it must also have a temperature. (The energy E of a static black hole just sitting there is simply its mass M, according to $E = mc^2$ as Einstein told us, with c set to 1 as is now customary.) So once we know what the entropy S of a black hole is, we could immediately differentiate to obtain the temperature T of the black hole.[30] An object with a temperature necessarily radiates. That's its way of sharing energy with its surrounding space. It is of course well known by now that Stephen Hawking worked out the details of how this radiation came about. Advice to physics students, and to the readers, "Take the lawfulness of creation as a matter of the utmost seriousness!"

Incidentally, before Beckenstein, people thought that black hole has infinite entropy, since everything and the kitchen sink could fall into it. According to the formula $T = \frac{dE}{dS}$, the temperature should then be zero, which means no radiation, hence totally black.

Since the work of Beckenstein and Hawking, entropy[31] has taken on an increasingly central role in physics. In particular, the entropy of a black hole is widely believed to be the key toward a deeper understanding of quantum gravity. The connection to information has also flourished through the enormous interest and hope invested in the possibility of quantum computers, to the extent that the two subjects, black holes and quantum computers, are increasingly intertwined.[†††]

Spreads nonsense far and wide and forever perverts brains

Boltzmann's refulgence continues to envelop our understanding of the world while his leading attacker Ernst Mach has become essentially worthless as far as physics nowadays is concerned. (Calm down. Just my opinion.) Note that while we frame our discussion of Boltzmann's work in terms of atoms and molecules and energy levels, he lived long before the atomistic picture of nature was established, and of course, at that time, quantum physics had not been even dreamed of by anybody. Not being able to convince his adversaries of his ideas no doubt led to his tremendous anger and frustration.[32]

Physicists venerate Boltzmann for many reasons and mourn his suicide not long before what should have been vindication for his ideas and the rectification of the abuse he suffered for them. I for one place Boltzmann in the highest pantheon of physicists, in the company of Newton, Maxwell, Einstein, and the like, and I believe that many physicists share my opinion. I admire him not only for his profound contribution to physics, but also for his fearless and fiery dismissal of philosophers.

In an ironic twist of fate, Mach, who held the chair of physics and philosophy at the University of Vienna, suffered a stroke. (For what it is worth, I might mention that Einstein referred to Mach as "a miserable philosopher" and Planck called him "a false prophet" leading physics astray.) Poor Boltzmann, though slighted by not holding a chair, had to give his nemesis's lectures on philosophy. The brilliant physicist made a valiant attempt to read up on various leading philosophers. I will let Boltzmann describe his efforts: "I first turn to Hegel; but, oh! what obscure vacuous balderdash did I find there. My unlucky star then ushered me from Hegel

---

[†††]Indeed, the study of the information content of black holes offers perhaps one of the few substantive handles we have on quantum gravity.

to Schopenhauer." After studying Schopenhauer's work, Boltzmann proposed a lecture with the title[33] "Demonstrations That Schopenhauer Is an
Insipid and Ignorant Philosopher Who, by Ceaselessly Propagating Hollow Twaddle, Spreads Nonsense Far and Wide and Forever Perverts Brains
from Top to Bottom." Academia being what it was (and still is), Boltzmann was compelled to drop this title and use the much shorter title "On
a Thesis of Schopenhauer's" instead.[34]

Alas, more than a century later, Hegel and Schopenhauer continue to rot
the brains of impressionable young people. The only people who benefited
from studying Schopenhauer are those lucky few who secured chairs at
elite universities, and live out their lives surrounded by adoring undergraduates, an infinitesimal fraction of whom would then go on to hold chairs
thus perpetuating the cycle. No doubt, however, that at a typical American
university, those who have heard of Schopenhauer vastly outnumber those
who have heard of Boltzmann. Schopenhauer's revenge?

It would perhaps be wise for me to close this long chapter with a
light anecdote, but given Boltzmann's tragic end, I could hardly do so.
Instead, I will tell you that, among physics students, the opening of a classic textbook[35] on thermal physics has achieved almost legendary status.[‡‡‡]
The book starts thus: "Ludwig Boltzmann, who spent much of his life
studying statistical mechanics, died in 1906, by his own hand. Paul Ehrenfest, carrying on his work, died similarly in 1933. Now it is our turn to
study statistical mechanics. Perhaps it will be wise to approach the subject
cautiously."

I am simply passing on Goodstein's advice.[§§§]

Appendix: Derivation of Shannon's formula

Take the situation described in the fable, with the probability that the
prospective groom is handsome, ordinary looking, or ugly given by $p =$
0.1, 0.6, and 0.3 respectively. To be concrete, let us suppose that the
princess bride gives her spy a piece of paper with 100 boxes printed on
it, with 10 colored green, 60 colored yellow, and 30 colored red (like traffic lights). The spy is to look at the groom and mark one of the boxes. We

---

[‡‡‡] Indeed, the opening seemed so far from the norm of physics textbooks that it even attracted
the attention of Snopes, a leading website devoted to combating fake news: https://www.snopes
.com/fact-check/science-textbook-gloomy-intro/

[§§§] Another curious fact is that Ehrenfest was Boltzmann's student.

are assuming that the looks of the available men in the village could be divided into 100 types, so that the probabilities of belonging to each one of these types are all equal. (If 100 is not large enough, use 1,000, and so on: this is known as fine graining.)

So, 60 boxes are colored yellow and labeled as ordinary looking; in other words, there are 60 ways of looking ordinary. No ordering or ranking among these 60 is implied. Indeed, as was just said, we assume the opposite: each of the 60 types is equally likely, so that we could use our earlier formula $I = \log w$ for w equally likely choices. If the message to the bride comes back saying ordinary, then the information gained equals $\log 60$. Similarly, 10 boxes are labeled as handsome, and 30 as ugly, again without any sense of ranking within each color.

The spy returns and hands the princess the piece of paper in an envelope. Let us denote the information contained therein by $I(\{0.1, 0.6, 0.3\})$. We will try to determine what that equals.

The princess now looks at the paper. Suppose a yellow box is marked. By seeing which box is marked, the princess gains an amount of information equal to $\log 60$. But the probability that the marked box is yellow is 0.6, so we need to multiply the information gained by 0.6, thus obtaining $0.6 \log 60$ for the expected amount of information gained. Similarly, $0.1 \log 10$ or $0.3 \log 30$ if a green or red box is marked. So in total, we obtain $0.1 \log 10 + 0.6 \log 60 + 0.3 \log 30$.

But at this point, we see that one of the 100 equally likely boxes is marked. So the bottom line of this two-step process is that we have gained an amount of information equal to $\log 100$. Thus, $I(\{0.1, 0.6, 0.3\}) + (0.1 \log 10 + 0.6 \log 60 + 0.3 \log 30) = \log 100$, and we obtain an equation for $I(\{0.1, 0.6, 0.3\})$, namely

$$I(\{0.1, 0.6, 0.3\}) = \log 100 - (0.1 \log 10 + 0.6 \log 60 + 0.3 \log 30)$$

$$= (0.1 \log 100 + 0.6 \log 100 + 0.3 \log 100)$$

$$- (0.1 \log 10 + 0.6 \log 60 + 0.3 \log 30)$$

$$= -(0.1 \log 0.1 + 0.6 \log 0.6 + 0.3 \log 0.3)$$

(Here we have used the property of the logarithm, for example, $\log 100 - \log 60 = \log(100/60) = \log 1/0.6 = -\log 0.6$.)

Evidently, we could generalize from three categories to however many categories we like. We have thus derived Shannon's formula cited in the text:

$$I(\{p_1, p_2, p_3, \cdots\}) = -(p_1 \log p_1 + p_2 \log p_2 + p_3 \log p_3 + \cdots)$$

What if the probabilities are known to more significant figures, for example, $p_1 = 0.089$, $p_2 = 0.605$, $p_3 = 0.306$? No problem. Simply have 1,000 boxes on that piece of paper given to the spy, with 89 boxes colored green, 605 colored yellow, and 306 colored red. But then a mathematician would point out that this derivation works only for probabilities given by rational numbers. Well, most physicists would be satisfied and totally believe that the result actually holds true for probabilities given by any real numbers between 0 and 1. Indeed, for any real number, you could write down a rational number as close to that real number as one would like. See chapter 8 for the difference between mathematicians and physicists.

Notes

[1] Students asked to rank the four core courses of the undergraduate curriculum in increasing order of difficulty would typically say classical mechanics, thermal physics, electromagnetism, and quantum mechanics. However, since quantum mechanics is linear, it is easily learned once one has accepted quantum weirdness as contrary to common sense. My experience is that when professors are asked to rank, the typical order might be quantum mechanics, electromagnetism, classical mechanics, and then thermal physics. Classical mechanics, in contrast to quantum mechanics and electromagnetism, is nonlinear, and so contains some exquisitely difficult problems to solve. Thermal physics is full of subtle concepts such as temperature and entropy, and may in fact be the most challenging to master. Physics professors are asked to fill in some kind of teaching preferences for the next academic year, and people are typically clamoring and fighting to teach undergraduate quantum mechanics, by far the easiest course to teach.

[2] To appreciate how confusing thermal physics could be, you need only read a review in *Physics Today* by an expert savaging Feynman's book on the subject.

[3] I like to mention the sad story (http://en.wikipedia.org/wiki/JohnJamesWaterston?wprov =sfti1) of John Waterston, an obscure instructor of naval cadets during empire days in Bombay (now Mumbai) whom nobody in the elite physics circles in England at the time had ever heard of. He worked out in this remote imperial outpost all by himself what we now know as the kinetic theory of gases, even deducing that the combination PV/T is a constant characteristic of each gas. At a time when most physicists did not believe in atoms and molecules, Waterston submitted a paper for publication that, almost needless to say, was dismissed by some august authority as nonsense. Lord Rayleigh, recognizing Waterston's contribution long after his death, had this to say: "The history of [Waterston's] paper suggests that highly speculative investigations, especially by an unknown author, are best brought before the world through some other channel than a scientific society, which naturally hesitates to admit into its printed records matter of uncertain value. Perhaps one may go further, and say that a young author who believes himself capable of great things would usually do well to secure favourable recognition of the scientific world by work whose scope is limited, and whose value is easily judged, before embarking upon higher flights." I offer this as advice to those amateurs who bombard academic

physicists with claims that they had solved the universe. Better to say that you have solved this or that.

[4] Indeed some economists have used this distribution to study the distribution of wealth in capitalist societies. See Victor Yakovenko's talk at the KITP: "New Developments in Statistical Mechanics of Money, Income, and Wealth," https://online.kitp.ucsb.edu/online/colloq/yakovenko2/.

[5] When I say large, I mean large. The numbers in statistical physics textbooks, written down mostly to frighten the children, are humongous.

[6] Indeed, a leading textbook on thermal physics comes in two volumes: M. Kardar, *Statistical Physics of Particles* and *Statistical Physics of Fields*.

[7] Indeed, this criticism and others, unanswerable in the late 19th century, drove poor Boltzmann to his tragic end.

[8] This is an example of an instance in theoretical physics for which I feel, in the strongest possible terms, that rigor, instead of being merely window dressing (as described in chapter 8) is in fact, essential.

[9] For those who would like to know more, this is discussed in some detail in *QFT ASAP*.

[10] For a glimpse of his somewhat tormented life, see the short film *The Bit Player*.

[11] For instance, he was the only member of the Institute for Advanced Study in Princeton to resign after six months.

[12] There was earlier work in the 1920s by H. Nyquist and R. Hartley.

[13] The logarithm was said by some to be "one of the very greatest scientific discoveries that the world has seen." Incidentally, the (wonderful) slide rule is just a mechanical representation of the logarithm.

[14] So that the integers go like 0, 1, 10, 11, 100, 101, ...

[15] A high school student familiar with exponential and logarithm would know that for the sake of simplicity I am suppressing units of time and of money, interest rate, etc.

[16] See the biography of Dirac, by G. Farmelo, *The Strangest Man*, 2011.

[17] A few letters of recommendation have become legendary in theoretical physics. For example, Wolfgang Pauli's letter for Victor Weisskopf (later the luminary of theoretical physics at MIT): "I have nothing against this man."

[18] If you ever visit Yale University, be sure to pay your respect at Gibbs's grave. https://quantumfrontiers.com/2019/04/16/long-live-yales-cemetery/.

[19] Perhaps the following explicit demonstration would help some readers. Let $p = e^{-k}$ so that $p \to 0$ as $k \to +\infty$. Then $-\log p = k$ and so $-p \log p = ke^{-k}$, which goes to 0 exponentially fast.

[20] For a history and summary of this book, see https://en.wikipedia.org/wiki/What Is Life%3F.

[21] The term negentropy was later coined by the French physicist Léon Brillouin. https://en.wikipedia.org/wiki/Negentropy.

[22] See, for example, W. Bialek, *Biophysics: Searching for Principles*, Princeton University Press, 2012. See also W. Bialek and A. Zee, *Phys. Rev. Lett.* 58: 741, 1987; ibid. 61: 1512, 1988.

[23] What I am describing here is called a microcanonical ensemble. Physicists also deal with the canonical and the grand canonical ensembles.

[24] For those readers who want to see the math, I offer here an extremely concise derivation of the Boltzmann distribution. To go through this, the reader has to know Lagrange's approach to maximizing a function with constraints imposed, and also that $\frac{d}{dp} \log p = \frac{1}{p}$ for the natural logarithm.

Lagrange instructs us to multiply each constraint by a constant, now known as Lagrange multipliers, and add these to the function we wish to maximize. In other words, form, in the present context, the modified entropy

$$S_{\text{Lagrange}} = \sum_i (-p_i \log p_i) + a \sum_i p_i E_i + b \sum_i p_i$$

The constants a and b are the famous Lagrange multipliers. Lagrange told us to then maximize this modified entropy by the standard Newton-Leibniz method of differentiating with respect to p_j (with $j = 1, 2, \cdots$) and setting the result to 0:

$$\frac{\partial S_{\text{Lagrange}}}{\partial p_j} = -\log p_j - 1 + aE_j + b = 0$$

Remembering that $e^{\log p_j} = p_j$, we can solve this equation immediately and obtain

$$p_j = e^{-1+b}e^{aE_j} = \frac{1}{Z}e^{-\frac{E_j}{T}}$$

In that last equality, we have simply given new names to the two expressions involving Lagrange multipliers, namely $Z = e^{1-b}$ and $T = -\frac{1}{a}$. (Actually, we barely have to write these two equations down; the Lagrange multipliers a and b, having done their jobs, are replaced by Z and T.)

By simply plugging into the two constraints

$$\sum_i p_i = 1 \quad \text{and} \quad \sum_i p_i E_i = E$$

we then determine Z and T according to

$$Z = \sum_i e^{-\frac{E_j}{T}} \quad \text{and} \quad \frac{1}{Z}\sum_i E_i e^{-\frac{E_j}{T}} = E$$

As you can see, the temperature T is determined by the average energy E, but the relation between them is highly implicit.

[25] For those readers who would like to see how this emerges, I now sketch the calculation. For convenience, write $\beta = 1/T$. Then Boltzmann gave us $p_i = \frac{1}{Z(\beta)}e^{-\beta E_i}$ with $Z(\beta) = \sum_i e^{-\beta E_i}$. Hence the entropy $S = \sum_i(-p_i \log p_i) = \sum_i(-p_i(\log e^{-\beta E_i} - \log Z(\beta))) = \sum_i p_i(\beta E_i + \log Z(\beta)) = \beta E + \log Z(\beta)$.

Next, $E = \sum_i E_i e^{-\beta E_i}/Z(\beta) = -\frac{d}{d\beta}\log Z(\beta)$. As in the text, since E_i is fixed, we do not distinguish between $\frac{d}{d\beta}$ and $\frac{\partial}{\partial\beta}$. Differentiate to obtain $dE = -(\frac{d^2}{d\beta^2}\log Z(\beta))d\beta \equiv +\frac{1}{\beta}d\tilde{S}$. The last step is merely the definition $\tilde{S}$.

Our task is to show that $\tilde{S}$ is none other than S. To obtain $\tilde{S}$, we simply integrate $\frac{d\tilde{S}}{d\beta} = -\beta(\frac{d^2}{d\beta^2}\log Z(\beta))$. Thus, integrating by parts, we find

$$\tilde{S} = -\int^{\beta} d\beta'\, \beta'\left(\frac{d^2}{d\beta'^2}\log Z(\beta')\right)$$

$$= -\beta'\frac{d}{d\beta'}\log Z(\beta')\Big|^{\beta} + \int^{\beta} d\beta'\, \frac{d}{d\beta'}\log Z(\beta')$$

$$= -\beta\frac{d}{d\beta}\log Z(\beta) + \log Z(\beta) = +\beta E + \log Z(\beta) = S$$

In the last step, we recognize that we have obtained exactly S as given earlier. Thus, the relation $dE = +\frac{1}{\beta}d\tilde{S} = Td\tilde{S}$ we obtained earlier is precisely $dE = TdS$, as desired.

[26] To add to the confusion, let me mention something for the benefit of some readers. Beware, other readers might become even more confused. For a given system, we are given the function $E(S)$. (For a different system, we are given a different function $E(S)$, of course.) Simply differentiate to obtain $T = \frac{dE}{dS}$, that is, determine T as a function of E and S. Since E is a given function

of S, we can also think of T as a function of S. But then we can solve for S as a function of E and T, or as a function of T. Now plugging in, we obtain E as a function of T, that is, we obtain a function $E(T)$. For some systems, this function $E(T)$ could well be a mess, but for a container of gas in the ideal approximation (that is, with various unnecessary complications neglected), the function is remarkably simple, just $E(T) = \frac{3}{2}NT$, with N denoting the number of atoms in the gas. Given the simplicity of this relation, this is the system students typically learn first, thus furthering the impression that E and T are simply proportional to each other (which in general they are not).

[27]Strictly speaking, the symbols Q and dQ, used in some books to denote heat and the infinitesimal change in heat respectively, should not be allowed in thermal physics. By dQ, some people actually mean TdS, but this is not the d of something called Q. In better books, dQ is written with a slash through the d, meaning that it is not really a d.

[28]Incidentally, the quantity TdS is also sometimes called the heat transfer, but it is important to keep in mind that "heat" is not a fundamental concept in physics.

[29]J. A. Wheeler, *A Journey into Gravity and Spacetime*, Scientific American Library, W. H. Freeman, 1999, page 221.

[30]For more details, see, for example, *QFT Nut*, *GNut*, *G*, and *FbN*.

[31]I might mention here that the concept of entropy in the quantum world has been generalized by the Hungarian mathematician Alfréd Rényi (1921–1970) to what is known as the Rényi entropy $S^{(\alpha)} = (\log \sum_i p_i^{\alpha})/(1 - \alpha)$, where α is a real number. Readers with the appropriate mathematical background might want to show that in the limit $\alpha \to 1$ this becomes the standard expression for the entropy and information.

[32] Boltzmann's frustration stemmed partly from my rant in a footnote at the beginning of this chapter. Consider a gas molecule in 1-dimensional space (for simplicity). Its state is characterized by its position x in space and its momentum p. In classical physics, the variables x and p both take on continuous values, and so it is impossible to count these states, as any decent mathematician would insist. So in the 19th century, physicists such as Clausius and Gibbs had to rely on an artifice (known as a fudge these days) saying that x and p are actually discrete on some infinitesimal scales δx and δp, so that when we integrate over the position and momentum of the molecule we have to divide by the size $\delta x \delta p$ of a fictitious "cell," something like $\int dx dp / \delta x \delta p$. (Another example of the difference between mathematicians and physicists as explained in chapter 8. A physicist forges ahead even in the face of a mathematical impossibility—count the possible values of a continuous variable!) It is at this point that undergraduates like me might throw the textbook on classical statistical mechanics against the wall in disgust.

The tragedy of Boltzmann is that he lived in a transitional period in physics, when there was mounting evidence of the discrete character of matter (nevertheless vigorously denied by reactionaries such as Mach) but before the advent of quantum physics. The German physicist Otto Sackur and the Dutch prodigy Hugo Tetrode independently published (in 1911 and 1912 respectively, when Tetrode was seventeen) what is known as the Sackur-Tetrode equation, with the recognition that Planck's constant h provides the mysterious size $\delta x \delta p$ of the fictitious "cell" introduced in a fictitious subject known as classical statistical mechanics.

Incidentally, Tetrode belonged to a wealthy and prominent Dutch family dating back to the Middle Ages, and lived in a mansion in Amsterdam. After reading Tetrode's paper, Einstein and Ehrenfest went to see him, but were sent away by the housekeeper who said that the young master was busy calculating. (I thank S. Lamoreaux for telling me this story.)

Unhappily, both Sackur and Tetrode died relatively young, in their early thirties, respectively from an explosion while doing military research in 1914 and from tuberculosis. (Nowadays, some students are more likely to associate the word tetrode with transistors having four active terminals.)

[33] Another translation I came across reads "Proof That Schopenhauer Is a Stupid, Ignorant Philosophaster, Scribbling Nonsense and Dispensing Hollow Verbiage That Fundamentally and Forever Rots People's Brains." I do not know which translation is closer to the original. By the way, "philosophaster" is the title of an early 17th century play by R. Burton. I wonder if there is a 21st century word for the pompous philosophy professors who feed Hegel and Schopenhauer to the young innocents at various universities?

[34] Delivered in Vienna on January 21, 1905.

[35] D. L. Goodstein, *States of Matter*, Dover, 2014.

10
PHYSICS IS WHERE THE ACTION IS

From one instant in time to the next instant

That physics could be formulated in terms of an action is to me at once astonishing and mysterious.

First, recall how physics is usually formulated. Students start with the equation of motion $F = ma$: Newton revolutionized physics by asserting that force F produces acceleration a, not velocity, as I already mentioned in chapter 2. At each instant in time, acceleration tells us what the velocity will be at the next instant, while the velocity at this instant tells us where the particles will be at the next instant. The force at this new location could well be different, which specifies what the acceleration will be at the next instant. These steps repeat infinitesimally and infinitely, thus directing the particle where to go and how to get there.

This is known as the local or differential formulation of Newtonian mechanics. (In everyday language, the word local is associated with space, but in physics and mathematics, the notion is generalized to include time.) Mathematicians call this an initial value problem.

Maxwell's equations of motion for electromagnetism were formulated with the same instant to instant outlook, telling us what the electric and magnetic fields will be at the next instant given how they are varying in space and what the electric charges are doing at this instant.

Similarly, Einstein's theory of curved spacetime masquerading as gravity tells us how spacetime and the distribution of energy should change from one instant to the next. And in quantum mechanics, Schrödinger's equation tells us how the probability amplitude changes from one instant to the next. Even when we graduate into the relativistic domain of particle physics and quantum field theory, we still have equations of motion specifying how various fields should evolve in time.

This formulation of physics in terms of equations of motion, determining what will transpire at the next instant in terms of what is happening right now, makes total sense to the human mind, which seems to operate on the same principle of moving forward in time, living from moment to moment. Physical reality appears to evolve from instant to instant, passing in front of us like a film in progress.

From local to global

Thus, it may come as a total surprise to you that these local or differential equations of motion could also be formulated in terms of an action principle which states that the particle follows a path that extremizes a certain quantity known as the action. (The word "extremize" is a neologism that encompasses both maximize and minimize.) See figure 1, in which we contrast the two opposing "philosophies" of telling a particle how to move.

Instead of specifying the particle's position x_I and velocity v_I at some initial time t_I and following the particle as it moves from instant to instant, with the action principle we specify the particle's position x_I at some initial time t_I and its position x_F at some final time t_F. Rather than stating the force under whose influence the particle should move, we assign a real number called an action to every one of the possible paths leading from the initial location in spacetime x_I, t_I to the final location x_F, t_F.

Which path the particle actually follows is the one that extremizes the action.

We have what mathematicians call a boundary problem instead of an initial value problem. (The word "boundary" is normally associated with space, as in the boundary of a country, but in physics and mathematics, the notion is also generalized to include time, just like the word "local." The initial and final locations in spacetime, respectively x_I, t_I and x_F, t_F, are thought of as the boundaries, that is, endpoints, of the paths the particle may choose to follow.)

This is known as a global or integral formulation of Newtonian mechanics. Instead of living from moment to moment, we take a "holistic" view of our lives.

Here I would like to alert the reader to a potential confusion over terminology. Back in chapter 4 we mentioned Newton's distaste for action at a distance. In that context, the word action is used almost in its everyday sense, referring to the earth acting on the moon and on the apple, for instance. In this chapter, the word "action" denotes a quantity assigned

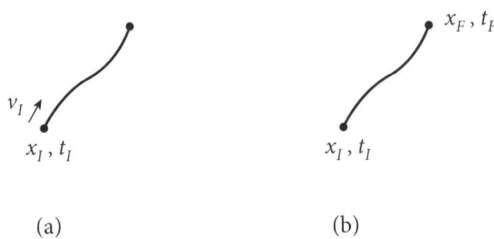

Figure 1. Two contrasting views of movement: (a) the local or differential view versus (b) the global or integral view. The two figures are the same but how we think of them differs. In (a), we start the particle off with an initial velocity v_I from an initial position x_I, and watch it move. If we start it off with a different initial velocity, it will follow a different path, of course, and in general will not get to x_F at time t_F. In (b), we "tell" the particle to get to x_F at time t_F starting from x_I at time t_I, and see which path it chooses. We do not specify the initial velocity v_I of the particle. Rather, the particle adjusts its initial velocity v_I so as to get to x_F at time t_F.

to each and every possible path a particle might "consider" following. (I haven't told you what that quantity is yet, and indeed it took physicists many decades to find it.)

Light in a hurry

The story of the action begins with the everyday observation that light beams form straight lines.

Easy to theorize why. Surely it is because light wants to cover the least amount of distance possible. Euclid could have told us. If we suppose that light moves at a fixed speed in air, we could say equivalently that light wants to get from here to there in the least amount of time.

The ancient Greeks[1] already understood that this principle also accounts for reflection off a mirror. See figure 2.

We see that this also mirrors another commonly observed phenomenon, namely that humans only covet what their neighbors have. Although the direct path from O to E is shorter than the path taken by the light bouncing off the mirror, the light going toward the mirror doesn't care. (Some light emitted by O of course does follow the direct path, and if the eyeball belongs to a champion basketball player with exceptional peripheral vision, it would also register that light.)

Light moves in a straight line, except when it doesn't

But light doesn't always travel in a straight line. Indeed, for you to read this book, eons of evolution have enabled your sainted mother to fashion a lens in your eyes, and to equip you with a brain and muscles to stretch and

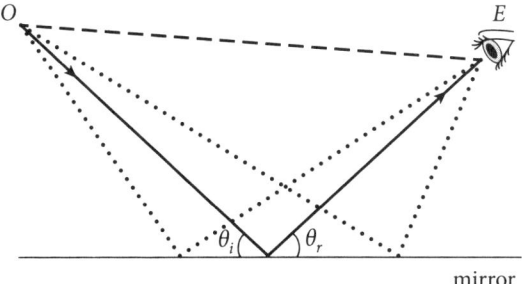

Figure 2. The path light actually takes (solid line) from an object O bouncing off a mirror and then to an observer's eyeball E is shorter than its neighboring paths (dotted lines). Simple geometry shows that for the shortest path the incident angle θ_i equals the reflection angle θ_r, as is observed in everyday life. The direct path (dashed line) from O to E actually provides an even shorter path.

squeeze that lens just so, to compel the light rays coming from the black marks on the page to bend and focus on your retina. The cycle may well close. Reading this book might enhance your advantages, perhaps boosting your chance of being selected, which in turn increases your prospects for reproducing yourself.

That light bends when it goes from one medium to another, known as refraction, is easily demonstrated in everyday life by sticking a spoon into a glass of water. Refraction[2] allowed Fermat to distinguish least time from least distance and thus formulate his eponymous least time principle. He had the insight that light would refract if its speed in water is less than that in air. See figure 3.

If it is distance that counts, light would go in a straight line (the dashed line in figure 3) from O to E. But if it is time that counts, and if light moves faster in air than in water, light would want to spend more time in air than in water, opting to bend as it goes from air to water and to follow a longer path (indicated by the solid line in figure 3) that takes less time.

Consider yet another path (the dotted line in figure 3). Light may try to gain time by spending still more time in air than in water, but then the extra distance makes that path a loser also.

Naturally, Fermat was savaged[3] by some of his contemporaries. "Who tells light which path to choose?"

The vanishing puddle of water and other illusions

During roadtrips on hot summer days, especially along interminably straight deserted highways in the American West, you often see a car far

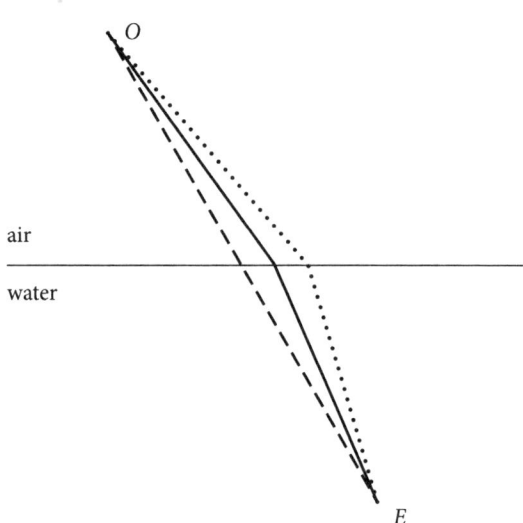

Figure 3. The path (solid line) light actually follows going from air into water is clearly not the path of least distance, but it takes less time than the two neighboring paths (the dashed line and the dotted line). Simple trigonometry and Fermat's principle of least time allow us to determine the refraction angle θ_r in terms of the incidence angle θ_i. (These angles are not marked in the figure to lessen clutter.) The resulting formula is known as Snell's law, an important discovery in its time only to be mocked much later by an economist.

away coming toward you going over a sheet of water, but yet when you get to that spot, the road surface is dry as bone. The physics behind this commonly observed mirage is explained in the caption of figure 4.

For the puddle of water mirage, the "covet only thy neighbors" principle also operates. The light taking the curved path skimming the road surface compares itself only with the light taking neighboring paths, not the light coming to your eyeball via the direct path.

You could think of your own experience of light bending. Here (figure 5) is a more dramatic version of the bent spoon in a glass of water.

The best deal for material particles

Fermat's least time principle for light led to a question for theoretical physicists. Is there a corresponding principle determining the path followed by material particles? Particles, in contrast to light, do not travel at a fixed speed. Least time clearly won't work. For instance, a falling object goes faster and faster, but a mass tied to a spring oscillates back and forth steadily, speeding up and slowing down depending on where it is in its cycle.

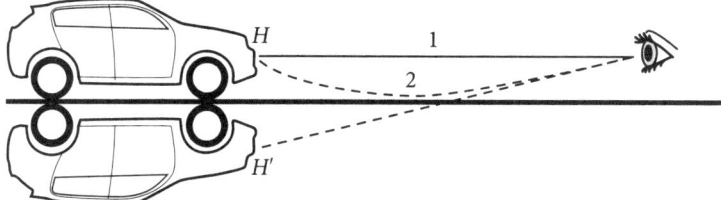

Figure 4. A common mirage on a straight highway on a hot dry day. Light moves faster in the hotter air near the road surface and so it refracts, following path 2. The "covet only thy neighbors" principle also operates. The light taking the curved path skimming the road surface compares itself only with the light taking neighboring paths, not the light coming to your eyeball via the direct path 1. Your marvelous brain interprets the refracted light as coming from the car's reflection on the road surface, thus deducing the presence of a sheet of water. Reproduced with permission from *Fearful Symmetry* (Princeton University Press, 1986).

Figure 5. Physics is phun! Permission to reproduce this photo granted by Bruce Illingworth, a distinguished teacher of high school physics in Mount Prospect, Illinois.

What are material particles trying to extremize, if anything?

Perhaps a word about terminology. In introductory physics, the word "particle" is used[4] for anything from billiard balls to the planets, as long as its "inherent properties," such as its size and rotation, are either irrelevant or unimportant.

Here I added "material" to emphasize that I am not talking about the "corpuscles of light" that Newton had theorized. (The irony is that in advanced physics, such as quantum field theory, light is again described in terms of particles, namely photons. The immortal Isaac was wrong and then correct yet once again, at least in some sense.)

Lagrange invented the Lagrangian

The race for the analog of a least time principle for material particles was on. After lots of confusion and priority fights among physicists, Leonhard Euler and his protégé Joseph Louis, Comte de Lagrange,[5,6] finally answered this question, some thirty years after Newton's death.

To explain how, allow me to speak anthropomorphically for a moment. At every instant, a material particle calculates a number $L(t)$, known to physicists as the Lagrangian (naturally!), and then adds up these numbers from start to finish, that is, integrates $L(t)$ over time t, for the readers who know some calculus. Those readers know that the sum of all these numbers is known as the integral $S = \int_0^T dt L(t)$, with the action traditionally denoted by S. (We are not going to actually calculate anything using calculus, so relax, and think of this as notation if you like.) That the action is an integral over time is plausible, in analogy with the travel time T of light in Fermat's least time principle, since T is just an integral of 1 over time: $\int_0^T dt\, 1 = T$.

For each path, we compute the action S. In other words, every possible path is labeled by a number, the action. The particle chooses the path that extremizes its action.

But what could $L(t)$ be?

Kinetic and potential energy

Before telling you what $L(t)$ turns out to be, I have to remind you of kinetic and potential energy in Newtonian mechanics (which in fact we have encountered before, for instance in chapter 5).

The energy of motion, namely the kinetic energy K of an object of mass m moving at velocity v, makes total intuitive sense. The more massive the object and the faster it moves, the larger its kinetic energy. In fact, $K = \frac{1}{2}mv^2$. Seems reasonable.[7]

Potential energy, denoted by V, is stored energy that could be released in the future, as in "that youngster has potential." You open a closet that you have not cleaned out for twenty years and a heavy box falls off the top shelf and smashes your toes. The work you performed twenty years ago lifting that box to the top shelf and leaving it there has been stored as potential energy. For twenty years, conservation of energy guarantees that not a bit of the stored energy has been lost, and now, in an instant, faster than you could get out of the way, it is all converted to kinetic energy. When we go downhill skiing, we pay the lift operator to provide us with lots of potential energy that we then convert to kinetic energy as rapidly as we dare.

As another example, a mass tied to a stretched spring has potential energy. When we release it, the potential energy stored in the spring gets converted to kinetic energy, and as the spring becomes compressed, that kinetic energy is converted back to potential energy. The mass oscillates, and kinetic energy and potential energy are converted to each other back and forth, while their sum, the total energy $E = K(t) + V(t)$, stays constant. (I emphasize here that K and V both depend on time t, while E does not.)

The variation of the potential energy from place to place gives rise to force. Perhaps even better: the spatial variation of the potential energy could be interpreted as a force. Historically, force came before potential energy, but mathematically, potential energy holds an advantage over force.[8]

The least action principle

In the blinding glare of hindsight, we see that L could only[9] be some combination of the kinetic energy K and the potential energy V. The sum, $K + V = E$, is already taken so to speak: it is the conserved total energy which, as a constant, cannot serve[10] as L. The only possibility is the difference $K - V$! The Lagrangian that physicists struggled to find for over a century is simply the kinetic energy minus the potential energy $L(t) = K(t) - V(t)$. Can't get any simpler than that! Again, hindsight[11] is always easy. Note that L, K, V all vary with time.

So, to summarize, for each path we calculate the action[12] $S = \int_0^T dt L(t) = \int_0^T dt (K(t) - V(t))$. Find the path that extremizes S. Out of all possible paths, this extremal path is the one the particle follows. Just as light wants to get there as soon as possible, every material particle in the universe wants to get the best deal, as measured by its action.

Remarkably, the path that wins is precisely the one that satisfies Newton's $F = ma$. Thus, the integral or global formulation produces the differential or local formulation, and so the two are equivalent.[13]

Since time immemorial, humans nearing the end of their lives have wondered how things could have gone better had they followed different paths. Human lives are of course infinitely more complicated than any system in physics we care to study. Different people may prefer to optimize different assets, anything from monetary to spiritual.[14] Furthermore, the analogy with physics fails because most people cannot predetermine the place and date of their demise. As a distinguished French physicist[15] once said to me, "You could only optimize your life locally, not globally." (As I have explained, the word "locally" is used in everyday life only for space, but in physics circles, it is understood to refer to spacetime.)

The action is a functional of the path

An "almost" unnecessary digression into mathematics, for those readers who may want to learn to talk the talk.

For an ordinary function $x(t)$, you feed in a number, t, and it gives you another number, $x(t)$. But here we generalize to one level higher. We feed in a path, which is described by a function $x(t)$, and ask for a number equal to the action associated that path. Mathematicians call the action S a functional of the function $x(t)$. You feed a number into a function to get a number, but you feed a function into a functional to get a number. Functional is to function as function is to number. (Yikes, what an ugly word!)

To find an extremum of a functional, given a function, you feed in its various neighboring functions, and ask if the functional, the action in the case of physics, goes up or goes down. This subject, essentially invented by Euler and Lagrange, is called variational calculus. As you can see, it's nothing to be frightened of; just one baby step higher than ordinary calculus. Conceptually, it is still based on Newton's and Leibniz's crucial idea. Vary the functional to see if you could do better. Same in life: we constantly vary, if only in our minds, to see if we could do better.

The action principle in action

Our friend the couch potato will give us our first hint of how least action works. Consider an object in the absence of an external force so that $V = 0$. You are supposed to get from x_I to x_F in time T.

The simplest example is when $x_F = x_I$: your destination is your starting point. Then of course the best strategy is to just lie there and not move, so that your kinetic energy K stays zero. Then the action S is zilch, nada. Any movement would generate kinetic energy and therefore increase the action. A body at rest stays at rest unless acted upon by an external force. Voilà! Newton's law of inertia.

"Too easy," you say. "The surprising part of Newton's mechanics, with which he bashes Aristotle to a pulp, is that, in the absence of an external force, a moving object will continue to move at the same velocity." (Recall the cursing peasant in chapter 2.) Let's check that this strategy, or path in spacetime, indeed leads to the smallest action possible. In order for the object to get from x_I to x_F in time T moving at a constant velocity u, that velocity has to be $u = (x_F - x_I)/T$. Thus, the action equals $S = \int_0^T dt K(t) = \int_0^T dt \frac{1}{2}mu^2 = \frac{1}{2}mu^2 T$.

Fine. But instead, suppose the object moves like some so-called jackrabbit drivers on the highway, sometimes going really fast, and sometimes really slow. To simplify the math, consider the simplest comparison case in which the velocity equals $u + w$ half the time and $u - w$ the other half (in order to maintain the necessary average velocity: $\frac{1}{2}((u + w) + (u - w))$ $= u$). Then, instead of u^2 in the action we just wrote down, we would have $\frac{1}{2}((u + w)^2 + (u - w)^2) = u^2 + w^2 \geq u^2$, which we see by using junior high school level algebra.[16] Thus, jackrabbit driving always leads to a larger action. As car owners' manuals remind us, in order to minimize wear and tear and to maximize gas mileage, cruise along at a constant velocity. We have obtained Newton's law of inertial motion.

The action principle in the presence of a force

But what if there is an external force, such as gravity? Then moving with a constant velocity is no longer the best deal. A falling object, say a cannonball dropped from the tower of Pisa, starts out slow, before gathering speed and hitting the ground. That's Galileo's law of acceleration of course. It follows path (a) in figure 6.

Let's try to understand what we observe as the cannonball's attempt to minimize its action. By staying at a high altitude where the potential energy is high for "as long as possible," it maximizes its potential energy V and thus lowers* the action $S = \int dt(K(t) - V(t))$. But then it has to rush at the end to get to the ground in the allotted time, and thus pays the price of a larger kinetic energy K. The actual path is a compromise between these two "desires."

First falling fast and then slowing down to meet the floor (path (c) in figure 6) is definitely a losing strategy. The action principle explains why it

---

*Because of the minus sign.

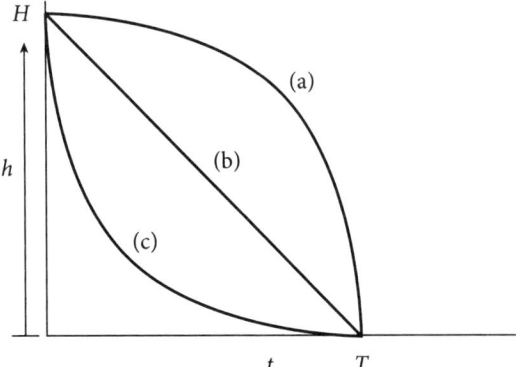

Figure 6. Which of the three paths, (a), (b), and (c), does a falling object choose? In (a), it starts out slow but then goes faster and faster. In (b) it falls at a constant speed. In (c) it rushes off and then slows down. From everyday experience, we know that (a) is the chosen path in space and time. Time t is plotted along the horizontal axis, and height h along the vertical axis. The falling object starts at height H and reaches the ground in time T.

would be a miracle to see a falling object first rushing off and then slowing down.

Driven by aesthetic dissatisfaction

It is an enormous pity that most people (and sadly, even most physics majors[17]) have never heard of the integral or global formulation and the action principle. Richard Feynman recalled that when he was bored out of his skull in high school physics class, his teacher, the legendary Mr. Bader,[18] showed him the action principle. That the particle somehow chooses its path in space and time blew the young man's mind. I believe that many physicists in my circle of acquaintances had a similar, almost mystical, experience upon first exposure to the action principle in Newtonian mechanics, and later, even more so, to the path integral formulation of quantum physics (which we will soon get to).

Cartoonists and others sometimes get the wrong impression that theoretical physicists are all busily calculating, covering entire blackboards with arcane formulas, as I have said repeatedly. In fact, theoretical physics at a certain level is not about getting results in agreement with experiments, but about understanding more clearly and deeply how the universe is put together. Of course, from day to day, much of theoretical physics is concerned with figuring out why a certain piece of metal has the electrical resistance it has or why a certain particle lives for such and such an amount of time before it decays. But the truly profound advance has often come when people ask seemingly useless questions such as why there is a least time principle for light but not something similar for material particles.

This same Feynman is also known in the physics community for likening theoretical physics to sex, as I've already mentioned in the prologue. Sure, sex sometimes produces useful results, but most of the time that is not why we do it, he said. But even Feynman sounded apologetic when he talked about the action principle in his famous giant red book.[19] He emphasized that most students could skip that chapter and that later chapters would not depend on it.

In my textbooks I like to digress and talk about a bright young physicist in a galaxy far far away. On this occasion, he or she or it might ask how come there are two separate extremal principles, one for light and one for material particles? Many physicists couldn't care less how many principles there are, but I must confess that I was troubled when I was a student. I now know that with the advent of Einstein's special relativity we could indeed combine these two principles into one.[20] Clear also that nonrelativistic physicists would be totally impotent trying to unify these two principles: the concept of K and V are alien to photons.

Indeed, our extragalactic bright young guy, driven purely by "aesthetics" and "desire to unify" (chapter 7), could, if he tried hard enough, manage to unify these two principles into one and discover special relativity in the process. It didn't happen this way on earth but it could well have. Unfortunately, in our present society and in our present utilitarian funding climate, a bright young guy doing this sort of thing is extremely unlikely to obtain a position at a prestigious university. The bean counters care about chalking up results, not principles. The equivalent of a young Lagrange these days may well not flourish.

Duality

That the same piece of physics could be described by different, even apparently contrarian, formulations—a meta-phenomenon known as duality—is both intriguing and profitable: the relevant physics often turns out to be much easier to understand in one formulation than in another. In our times, duality has been exploited intensively by string theorists. In particular, it turns out, mysteriously enough, that several apparently different string theories are related by duality.

The concept of duality goes back to electromagnetism: in the absence of electric charges and currents, Maxwell's equations are unchanged upon rotating the electric and the magnetic fields into each other.[21] More recently, it has popped up in statistical physics and in quantum field theory.[22]

Don't bother me with your alternative formulation

> You could validly argue that the minimum formulation is neat, but really no better than the other formulation. . . . The least time principle tells you how to formulate behavior under such conditions and the memorizing of Snell's law about angles does not. Who can doubt which is the better scientific explanation?
>   —Paul Samuelson, Nobel lecture for economics, 1970

Alternative formulations of known physics are sometimes referred to derogatorily as "mere formalism." If the action principle merely reproduces the equation of motion, why should practical minded physicists care? And indeed, most of them don't. They belong to the school[23] of "Don't bother me with your alternative formulation; I'm busy calculating." Ironically, a Nobel certified economist had to disabuse these practical types. If the least action principle leads to the equation of motion, then why not just start with the equation of motion in the first place, they say.

Exactly. Most of the time, why bother? But physics has a way of surprising us. Most alternative formulations wilt and fade away, but some have led to new insights, and a few may even become indispensable as new developments arise. The action principle is one of the most prominent examples historically. The practical minded people were totally wrong.

That new development was quantum physics.

Electromagnetism and the Lagrangian density

Before I get to quantum physics, I have to backtrack a bit. In mid-19th century, Maxwell synthesized all that was known and showed conclusively that light is in fact an electromagnetic wave. I consider, as would most physicists, Maxwell's theory of electromagnetism, as embodied in his equations governing how electric and magnetic fields varying in space and in time generate each other, as one of the greatest hits of physics.

Let us see how Maxwell's equations determine the refraction of light as it enters from one medium to another, such as from air into water. Inside a medium like water, the electromagnetic fields in light are, as to be expected, affected by the presence of water molecules, which after all are

next morning, Feynman had already written out in detail on the blackboard what is now known as the Dirac-Feynman path integral formulation of quantum physics.

A preternatural sense of the quantum physics

With this seldom taught Dirac-Feynman path integral formulation of quantum physics, the action takes on the starring role! The action formulation of classical physics turns out to be tailor made for the quantum world. Eerily, it almost seems that in 1760 Euler and Lagrange had a preternatural sense of the quantum physics to come.

Consider a particle moving from an initial location in spacetime x_I, t_I to a final location x_F, t_F. As explained in this chapter, a real number called an action is assigned to every one of the infinitely many possible paths. In classical physics, the particle "chooses" the path that extremizes the action.

What happens when we "switch on" the quantum? First off, you may recall that, contrary to the impression some websites might convey, quantum physicists do not traffic in probability, but in probability amplitude. They add up probability amplitudes, which are given by complex numbers, and then take the absolute square to obtain the probability, as was explained in chapter 3.

Within the path integral formulation then, we have to assign a probability amplitude to each of the infinitely many possible paths, and then we have to add them all up to find the probability amplitude that the particle will arrive at the position x_F at time t_F.

The difference between classical physics and quantum physics is thus summarized in this table:

classical physics	find the path that extremizes the action
quantum physics	add up the probability amplitudes of all possible paths

The fundamental postulate of quantum physics

But what probability amplitude are we supposed to assign to each path? For sure, this is not something for us humans to decide. Nature decides, and we accept that. My exposition strategy here is to ask you, yes you, to come up with the fundamental postulate of quantum physics.

There are some constraints of what probability amplitude we could assign to each path. We must recover the classical world as an approximation to the quantum world; after all, we live in the classical world, not the quantum world. There are not that many possibilities; indeed in hindsight, essentially only one. The probability amplitude, call it $\mathcal{A}$, assigned to each path must be characteristic of that path. What could $\mathcal{A}$ possibly depend on?

You could almost start to guess the answer, because I have carefully set you up. Guess what determines the probability amplitude for each path. Let's be clear that I'm not asking you to actually determine the probability amplitude. I am merely asking you which property characteristic of each path could possibly determine the probability amplitude.

Well, try the what-else-can-it-be method. By the way, when all else fails, the what-else-can-it-be method often works surprisingly well in theoretical physics. Some[28] regard it as the "last resort of scoundrels."

Actually, guessing that the action determines the probability amplitude is fairly close to what Dirac said.[29] Indeed, the probability amplitude has to be determined by some quantity that plays an important role in classical physics and that is characteristic of each path. The action S associated with that path is the only quantity[30] we know of that fills the bill.

To summarize, the probability amplitude $\mathcal{A}$ must depend only on S. So, here is our guess.

The fundamental postulate of quantum physics:
The probability amplitude $\mathcal{A}$ of a particle following a given path is determined by the action S associated with that path in classical physics.

To go from the classical world to the quantum world, we have to somehow construct the complex number $\mathcal{A}$ out of the real number S. Recall from chapter 3 that a complex number could be visualized as an arrow in a plane, that is, as a 2-dimensional vector, completely characterized by the length of the arrow and by the angle the arrow makes with respect to some reference direction.[31] Well, here comes the heart of quantum physics, so to speak. Trumpets!

The probability amplitude $\mathcal{A}$ assigned to each path is an arrow with length 1 and angle equal to $S(\text{path})/\hbar$, namely the classical action S of that path divided by the constant $\hbar$ introduced by Planck to kickstart quantum physics. See figure 7.[32]

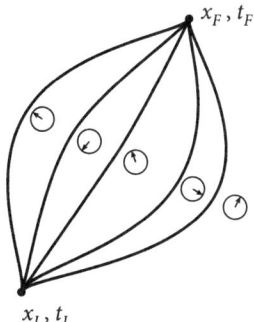

Figure 7. Assigned to each path is an arrow pointing in a direction determined by the classical action of that path.

Two important strands in our narrative have now come together. (1) Euler and Lagrange told us that the action S has dimension of energy multiplied by time, as indicated by our example of a classical particle $S = \int dt(K - V)$. (2) Planck's constant $\hbar$ has the dimension of energy multiplied by time, as was shown in chapter 3. Thus, S divided by $\hbar$ is a pure number, without any dimension. Before Planck came along, it would be utter nonsense to say that some angle is equal to the action S, because angle is a dimensionless quantity while S is not.

To summarize, in quantum physics the angle of the probability amplitude, or the direction the probability amplitude points in, is equal to the action of Euler and Lagrange divided by Planck's constant. Three giant intellects!

Quantum versus classical physics at a glance: circle versus line

Time for an executive summary.

	Associated with each path	
classical physics	action	a point on a line
quantum physics	probability amplitude	a point on a circle

Since the action S(path) ranges from $-\infty$ to $+\infty$, we could associate it with a point on the infinitely long real line. In contrast, since an arrow of length 1 defines a point on the circle with radius 1, the probability amplitude $\mathcal{A}$ "lives" on a circle. I like to say that when we wind the line around the circle we go from classical physics to quantum physics.

The Dirac-Feynman path integral formulation of quantum physics

Now that you have more or less grasped how the Dirac-Feynman path integral formulation of quantum physics works, I could go on and tell you an enormous amount[33] more about it. But I will restrict myself here to two remarks.

Those readers who have heard of quantum mechanics formulated in terms of the Schrödinger wave function might be getting a bit anxious. Where is the wave function in the path integral formulation? (Or, in the popular lingo of the internet, "Where is the darn cat?") No need to worry. The wave function and the Schrödinger equation that describes it simply pop out of the path integral.[34] (Of course, if the Schrödinger formulation didn't pop out, nobody would be talking about the path integral formulation, least of all greats like Dirac and Feynman.)

Perhaps even more anxiety inducing, certainly of more immediate and personal concern to many readers, is how this beloved classical world we were born into could emerge from such a bizarre quantum world where common sense ceases to reign.

Look at the two tables on pages 283 and 284, which should strike anyone at first sight that the classical and the quantum world operate on entirely different principles. In one, everybody chooses an extremal path to follow. In the other, the probability of getting from here to there is determined by a totally implausible and bizarre recipe. You find all paths leading from here to there, associate a probability amplitude with each path, add these all up, and then absolute square to find the probability you would actually get there. Could anybody honestly admit that this is not bizarre? Theoretical physics attracts many precisely because it is bizarro to the max!

How does our nice classical world emerge from this bizarre quantum world?

Emergence of the classical world from a bizarre quantum world where common sense ceases to reign

To give birth to the classical world, Nature has a very clever trick up her sleeve. The secret lies in how arrows are added together!

Recall from chapter 3 that in humanmade units, Planck's constant is measured to be about[35] 10^{-27} gram centimeter squared per second. This is fantastically small compared to anything in everyday experience, and thus explains why physicists discovered $\hbar$ and quantum physics only after they started exploring the microscopic world of atoms.

Each time you turn a page in this book, or blink, or utter a breath, not to mention any number of more strenuous activities, the action of your action is absolutely humongous compared to $\hbar$, as I had emphasized in chapter 3. Imagine the typical mass, distance, and time in grams, centimeters, and seconds involved in each of your colossal actions. If you move a 1-gram mass with a speed of 1 centimeter per second through a distance of 1 centimeter, your action amounts to a whopping $10^{27}\hbar$, or if you insist, 1,000,000,000,000,000,000,000,000,000 $\hbar$. A billion billion billion $\hbar$! The quantum world is like a foreign country with a fantastically weak currency unit $\hbar$, so that to buy a grain of rice you would have to fork over a billion billion billion local bills. We inhabit a classical world with massive lumbering lumps like ourselves.

Angle is a cyclic[36] quantity, that is, an angle[†] of $2\pi \simeq 6.28$ is equivalent to an angle of 0. Indeed, an angle of $2N\pi$, for any integer N no matter how large or how small, is still equivalent to an angle of 0. In other words, for your colossal action, you have to divide 10^{27} by 2π, and quantum physics only cares about the teeny remainder! In other words, if $10^{27} = 2N\pi + \varepsilon$ with $0 \leq \varepsilon < 2\pi$, then the corresponding probability amplitude does not depend on the value of N at all. Instead, the probability amplitude is simply an arrow of unit length pointing at an angle ε. You could see how remote quantum physics at its defining level is from human experience.

In the path integral, every path is surrounded by lots of neighboring paths, namely paths that differ from it slightly. (For ease of writing, call a path's neighboring paths simply its neighbors.) I remind you that classically a particle follows the path that extremizes the action, call it the extremal path.

For a generic path, its neighboring paths would in general have rather different actions. One of its neighbors might have a higher action, another a lower action. Then that path is clearly not an extremal path, because you could have either a higher or a lower action by following a neighboring path. In other words, those paths that are not favored by the extremal action principle are surrounded by paths whose actions differ substantially.

[†]Quantum theorists use radians rather than the Babylonian degrees. If you prefer, you could say that an angle of 360° is equivalent to an angle of 0°.

In contrast, an extremal path is surrounded by paths whose actions differ from the extremal actions only infinitesimally. As an analogy, imagine yourself hiking on a trail. When you are on a slope, a few steps in either direction would either increase or decrease your altitude. But when you reach the top or the bottom of the landscape, a few steps away and you will still be at more or less the same altitude. Readers acquainted with calculus would recognize this as the basic idea of Newton and Leibniz for finding the maximum or minimum of a function.[37]

Furthermore, as I had belabored, the action S(path) for any action performed in the classical world is enormous compared to $\hbar$: S(path) typically equals a gazillion $\hbar$. Alternatively, theoretical physicists often think of $\hbar$ as a dial we can adjust. As we turn $\hbar$ down to zero, we should see the quantum world transition into our good old classical world. A hypothetical universe with $\hbar = 0$ is a classical universe. Either way, something should happen when the ratio S(path)$/\hbar$ becomes huge.

Lots of little arrows pointing in the same direction can add up to a gigantic arrow

Remember that to obtain the probability amplitude for the particle to get from here to there we are supposed to add up the probability amplitudes of all the paths the particle could take. Perhaps you can start to see how Nature picks out the extremal path actually realized in classical physics. Her secret: "Lots of little arrows pointing in the same direction can add up to a gigantic arrow!"

Consider a path that is not extremal. Its neighbors have their little arrows pointing every which way, almost completely randomly. The angle in radians, or in degrees, the little arrows make with the vertical is huge compared to 2π, or 360, respectively, something of order 10^{27} or more. Remember, we don't care if you have gone around the circle 10^{27} times; it's the little bit that's left over that counts. The arrows are effectively pointing in every possible direction. All these arrows end up canceling each other. The total amplitude is to an extremely good approximation equal to zero. See figure 8(a).

In contrast, an extremal path is surrounded by neighbors whose little arrows are pretty much all pointing the same direction[‡] as that of

---

[‡]Incidentally, this is known as "stationary phase" to the mathematically knowledgeable, an approximation used in many areas of physics and engineering.

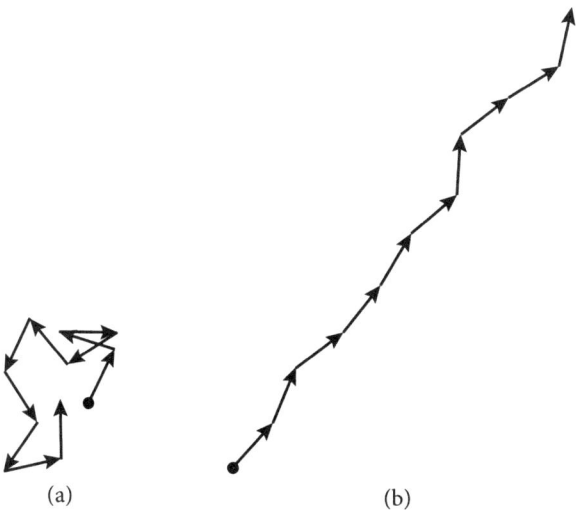

(a)   (b)

Figure 8. (a) Lots of little arrows pointing each and every way when summed up tend to cancel each other. (b) Lots of little arrows all pointing in the same direction add up to a big arrow. Keep in mind that in the path integral the number of little arrows is effectively infinite.

the extremal path. The little arrows add up to a gigantic arrow. See figure 8(b).

Another important point: You see that whether the action is maximized or minimized for our extremal path does not matter, only that it and its neighbors are all pointing in the same direction. I consider this to be one of the great triumphs of quantum physics, the explanation of which would otherwise pose a puzzle in classical physics, namely the puzzle of why the action is extremized, rather than minimized or maximized.[38] Just go ask a classical physicist why the action is sometimes minimized and sometimes maximized; he would be stumped.

That is more or less how Nature produces the world of the large and slow. Pick out the extremal path and forget about the rest.

Intertwining fields and the four fundamental interactions

As theoretical physics moved from quantum mechanics to quantum field theory, some of the familiar concepts abstracted from everyday experience, such as force, became less and less useful. Popular books on physics perhaps tend to give the opposite impression. You can talk, rather vaguely, of the forces between the eight different gluons of the strong interaction, but it is exceedingly awkward. Instead, we have intertwining quantum fields[39] interacting in ways totally alien to the world of billiard balls colliding. This started in the 1930s with the likes of Dirac and Fermi, gathered force during the late 1940s thanks to young hotshots like Feynman and Schwinger, and then really picked up momentum in the 1960s.

The four fundamental interactions, the strong, the weak, the electromagnetic, and gravity, with their actual or potential unification, as was described in chapter 7, have all been formulated in terms of the action principle. Sure, you can formulate quantum chromodynamics using the Schrödinger or Heisenberg approaches, and some people do, mostly to publish papers. Almost all of contemporary particle theory is developed using the Dirac-Feynman approach (employing various approximation schemes, such as Feynman diagrams, of course).

When people started speculating about strings underlying the universe, they didn't think about the forces between the different segments of a string, and they didn't write down the corresponding equation of motion. Instead, they[40] wrote down the action.

To me, that physics, particularly at the fundamental level, could be formulated in apparently quite different ways is both intriguing and mysterious.

The action is compact

Since you could always vary an action to obtain the equations of motion, the action could be thought of as containing the equations of motion tightly packed inside. The action, being much more compact, is significantly easier to remember than the equations of motion.

Undergraduates learning electromagnetism are shown four coupled partial differential equations, first written by Maxwell, describing how variations of the electric and the magnetic field in space and in time generate each other, as I already mentioned in chapter 4. As an undergraduate I always had trouble remembering this full set of rather complicated looking equations. But as a graduate student I had no trouble whatsoever remembering the action, which turned to be as simple[41] as can be. Once you know the action you could readily derive Maxwell's equations, and with some practice, quickly in your head.

Similarly, the action for Einstein gravity is extraordinarily compact, much more so than the ten Einstein field equations.[42] As was mentioned in chapter 6, Einstein struggled arduously for ten long years searching for the field equations. With the benefit of hindsight which of course he didn't have, he could, arguably, have found the action with much less effort.

A somewhat imperfect analogy, already hinted at earlier, would be some gizmo ordered online arriving packed tightly in a surprisingly small box. The box unpacks into numerous complicated looking pieces. But once unpacked, it is literally impossible to fit them all back into the box. Historically, the equations of motion for Newtonian mechanics, for Maxwell's electromagnetism, and for Einstein gravity were all discovered first. The corresponding action was discovered later, and in the case of Newtonian mechanics, Euler and Lagrange had to invent the concept of action. In our analogy, you come across a pile of pieces, but nothing even tells you that there is a box, let alone that those pieces could be neatly packed into a box. But starting in the second half of the 20th century, the action often precedes the equations of motion, as was the case for Yang-Mills theory and also for string theory.[43]

Symmetries are manifest and evident in the action

A tremendous advantage of the action formalism, which makes it the easy winner for fundamental physics since roughly the middle of the 20th century, is that the all important symmetries discussed in chapter 5 are manifest and evident in the action formulation, but less so in the equation of motion. Under the transformation corresponding to a particular symmetry, the action does not change: it is invariant.

As an example, suppose you want the relativistic action[§] of a point particle with mass m. Well, what is something pertaining to this particle in spacetime that is invariant under the Lorentz transformation? There exists only one quantity that fits the bill: the length of its path in spacetime. Being geometrical, the length by construction does not depend on the observer. (We're looking at you, the "truth is relative" philosophy professors.) To obtain a quantity with the dimension of an action, we have to multiply the length by a mass. That's it! So incredibly simple and easy to derive.

[§]There is a common and potentially serious misconception here: the adjective "relativistic" does not mean that the action holds only for a particle moving with a speed close to the speed of light c. It actually means that it holds regardless of speed, up to c, but also including the case of a particle sitting at rest. In other words, the word relativistic is meant to be inclusive rather than exclusive. I had already belabored this point back in chapter 4, but it is well worth repeating.

By the way, this provides one of my favorite definitions of mass. The mass m of a particle is its intrinsic quality allowing it to link geometry with physics.

By varying the action, we obtain the equations of motion, as Euler and Lagrange taught us. In all known cases, the equations are significantly more complicated looking and involved mathematically. For example, the action for electromagnetism is a lot simpler looking than Maxwell's equations (a secret that many professors try to hide from undergraduates!). But perhaps the most egregious example is the Einstein-Hilbert action versus Einstein's field equations, which took Einstein ten years to find, as I just mentioned. I believe that Einstein could have saved himself lots of grief and time if he had used the action principle from the start. Indeed, once Hilbert understood what Einstein was up to, he was able to write down the action in a matter of days, thus potentially scooping Einstein and provoking an extremely bitter fight.[44]

The embodiment of the symmetry in the equations of motion is significantly more involved and far less transparent than in the action. First, there is only one action but many equations of motion, for example, Maxwell's equations following from a compact action. Under a Lorentz transformation, each term in an equation transforms: none of the terms is invariant, and shouldn't be. Rather, the symmetry requires that each term in each equation transforms the same way, that the equations of motion are covariant.

This is one of the few places where it pays to be pedantic (recall what I said in the prologue): the action is invariant, the equations of motion are covariant. In all cases I know of, it is much easier to check that something is unchanged than to check that everything has changed but all in the same way by the same amount.

The reader does not need to know the mathematical details to see that this may be plausible: after all, the equations of motion were tightly packed inside the action, as in our admittedly imperfect analogy.

The action principle and unification

I promised in chapter 7 that I would explain what unification means in terms of the action principle. (I will be compelled to simplify and gloss over some details.) At some point in history, the action describing the three nongravitational interactions would have the form $S = S_{strong} + S_{electromagnetism} + S_{weak}$, that is, the sum of three separate terms. The action is a functional of the various fields we know about, the quark fields, the electron field, the neutron field, the gluon field, and so on and so forth. When we plug in any configuration of these fields in spacetime, the action produces a real number. (I am providing not even a bird's-eye view, but a view from the moon so to speak!)

The plus sign has precisely the meaning we learned way back when in elementary school. The separate pieces of the action are literally added together, and are separate. On a whim, we could cut out one of the pieces. You want a universe without a weak interaction? Fine, no problem. Here is the action for your universe: $S = S_{\text{strong}} + S_{\text{electromagnetism}}$. You want a universe with only the weak interaction? Also, no problem. Here: $S = S_{\text{weak}}$.

After electroweak unification, the action acquires the form $S = S_{\text{strong}} + S_{\text{electroweak}}$. The action $S_{\text{electroweak}}$ can no longer be written as the sum of two pieces: $S_{\text{electroweak}} \neq S_{\text{electromagnetism}} + S_{\text{weak}}$. Electroweak unification means that the electromagnetic and the weak interactions are now completely intertwined. And after grand unification, the action consists of just one single piece in which the three nongravitational interactions are intertwined into one single interaction: $S = S_{\text{grand unified interaction}}$. All of this would be extremely awkward to express using equations of motion, and in practice, essentially nobody I know in fundamental physics uses that language any more.

The action principle is discriminating

As I have said, physics advances on many fronts. Indeed, most of it (I am not sure how you quantify "most"; by the number of papers pouring out?) is concerned with understanding the multitude of what Einstein majestically dismissed as "this or that phenomenon," as was mentioned in the prologue, and proceeding more or less along Anderson's "more is different" philosophy (which nobody, certainly not I, disagrees with, as I said in chapter 7). Yet, some of physics is striving to know how the fundamental laws of the universe are put together. On this front, the action principle has totally dominated.

Students of physics are confronted with a bewildering array of laws, as I said in chapter 7, but they soon realize that these laws could be arranged in a hierarchy. In importance and applicability, they range from the analogs of local ordinances about when and where you are allowed to cross the street to some universal law declaring human rights. Take for example Newton's equation with a friction term added: $F = ma + \mu v$, or if you prefer, $F - \mu v = ma$. Here v denotes the velocity of the particle and μ the so-called coefficient of friction characterizing the contact between the object and the table (just to be definite) it is moving on. We know full well that friction is not a fundamental force but arises merely due to minute irregularities on the surfaces of the moving object and of the table. This kind of example occurs again and again in physics, and indeed there are people who devote their lives to studying dissipation. Often this is simply due to neglected effects[45] evident in classical physics. Interestingly, it is either awkward or impossible to formulate these so-called phenomenological equations of motion in terms of an action.

This brings us to a mysterious point. Given an action, Euler and Lagrange taught us how to vary it to obtain an equation of motion. You could always do this. In contrast, if somebody writes down an equation of motion and asks you to find an action whose variation would give that equation, in general it cannot be done. Readers familiar with calculus might be reminded of a similar phenomenon. Given a smooth function, no matter how complicated or nasty looking, we could always differentiate it to find its derivative. But in general you may not be able to go the other way. In other words, given a function, even some looking quite innocent, it could be essentially impossible to evaluate its indefinite integral (sometimes called the anti-derivative) analytically.[46]

Devoted to action

So it is quite amazing that when everything is properly taken into account, the fundamental laws of physics, and hence all of physics at least in principle, can be formulated in terms of the action principle.

This to me is a great mystery of physics.

Yes, I and many of my friends have an almost mystical devotion to the action principle. Of course, tomorrow somebody could come up with a drastically new formulation of gravity that cannot be written as an action principle, but that remains to be seen. I am waiting patiently. Nobody but nobody in the theoretical physics community would ever claim that physics is a finished piece of work.

An exciting development, spearheaded by Arkani-Hamed and his collaborators and others, is to start with the probability amplitudes for the scattering of particles, and try to fix these probability amplitudes by imposing various general requirements without referring to the action. In this so-called amplitude approach, even spacetime is not put in from the start, but emerges. Upon first hearing, this formulation sounds like black magic: physics is known to start by thinking about movement through space as time elapses. But with the quantum world undergirding our illusory classical world, it might make sense to treat probability amplitudes as more fundamental than even spacetime.[47]

From least time to action, from path integrals to quantum field theory, we have four hundred years of intellectual continuity in physics. For this, we are grateful to the likes of Lagrange and Hamilton, who in the eyes of their contemporaries were wasting their time not busily calculating.

Proudly born in the mid-18th century, with a pedigree going back to Fermat's least time principle in the mid-17th century, the action principle hardly needs me to sing its praise. Strangely, neither Maxwell nor Einstein used it in any essential way, even though they were aware of it. Indeed, the action principle was largely ignored by physicists till Dirac and Feynman gave it a central role in quantum physics in the mid-20th century. Ever since then, it has occupied the place of honor in the hearts of fundamental physicists. Curiously, even today the action formulation of quantum physics is normally carefully hidden from undergraduates, revealed only occasionally to an exceptionally bright kid, perhaps the way Mr. Bader showed it to Feynman in high school.

Notes

[1] Hero of Alexandria (AD ca. 10–70) did.

[2] A water wave coming into a beach also refracts, so that the front of the wave ends up parallel to the beach. Did you ever wonder about that? This is because the speed of a water wave decreases as the depth decreases. See *FbN*, page 279.

[3] The least time principle undoubtedly has a quasi-theological flavor, which unfortunately has caused much ink to be spilled.

[4] As was already used in this sense in the first section of this chapter.

[5] Lagrange was a teenage prodigy, but fell into a deep depression in his old age. Fortunately for him, the daughter of Lemonnier, an astronomer friend of Lagrange's, managed to cheer him up. Almost forty years younger than Lagrange, the young woman offered to marry him. Soon Lagrange was productive again. See L. Reimer and W. Reimer, *Mathematicians Are People, Too*, Dale Seymour Publications, 1994, page 88.

[6] Grange means barn in French. Strictly speaking, Lagrange was Italian by place of birth, but he was born before there was such a thing as Italy.

[7] This fact was also mentioned in chapter 6. Recall Einstein's derivation of $E = mc^2$?

[8] The potential energy at a given point in space, denoted by $V(\vec{x})$, is a single number, but the force at a given point in 3-dimensional space consists of three numbers, one for each of the three Cartesian directions. Much easier to keep track of one number than three.

[9] The possibilities are very limited. We have to use quantities that are universal for any particle moving in all possible ways. How about momentum? Well, no, it is a vector, while L is just a number. On the other hand, momentum squared does appear in $K = \frac{p^2}{2m}$.

[10] Because then the action S would be simply ET, and does not depend on any details of the actual motion.

[11] Staircase wit, *l'esprit d'escalier*, Treppenwitz, firing the cannon after the cavalry had already charged by you.

[12] For readers not so fluent with the notion of integration (it's just a fancy sum), an analogy might help. The computation of the action is similar to that done by an accountant determining the total profit of a business for any given production strategy. He subtracts the total cost of production from the gross income on a weekly basis and then sums this quantity over the 52 weeks in the fiscal year. The Lagrangian corresponds to the weekly profit, the action to the annual profit. The businessperson naturally tries to maximize the total profit by following the most advantageous strategy or history. An executive summary in a table:

business	gross receipts	cost of production	weekly profit	annual profit
physics	kinetic energy	potential energy	Lagrangian	action

[13] I could show those readers who know some calculus how that works, using a simplified example. Consider $J = \int dt (\frac{df}{dt})^2$. To determine the function $f(t)$ that extremizes J, we find out how the neighbors are doing. Look at a neighboring function $f(t) + \eta(t)$ where $\eta(t)$ is an infinitesimal function that vanishes at the endpoints of the integral. For this neighbor, $J_{neighbor} = \int dt (\frac{df}{dt} + \frac{d\eta}{dt})^2 = \int dt \{(\frac{df}{dt})^2 + 2\frac{d\eta}{dt}\frac{df}{dt} + (\frac{d\eta}{dt})^2\}$. Thus, if $f(t)$ satisfies the "equation of motion" $\frac{d^2f}{dt^2} = 0$, the change $(J_{neighbor} - J) = \int dt \{-2\eta\frac{d^2f}{dt^2} + (\frac{d\eta}{dt})^2\}$ equals $\int dt (\frac{d\eta}{dt})^2$, which is manifestly positive. (Here I have integrated by parts, remembering that $\eta(t)$ vanishes at the endpoints of the integral.) We conclude that, if $\frac{d^2f}{dt^2} = 0$, $J_{neighbor}$ is larger than J. Thus, J reaches its minimum if the "equation of motion" is satisfied. Lagrange leads to Newton. I will leave as an exercise for you to find the extremum of the more elaborate $J = \int dt \{(\frac{df}{dt})^2 - V(f)\}$. The "equation of motion" becomes $\frac{d^2f}{dt^2} = -\frac{1}{2}\frac{dV}{df}$.

[14] Another analogy that comes to mind is the Choose Your Own Adventure books that were a rage among children worldwide starting in the 1980s, which sadly were wiped out by video games. The difference again is that the endpoint is not fixed: early death if you choose to accept a mouthwatering pastry from a friendly vampire.

[15] By the way, this same French friend arranged for me to spend a year in Paris. Talk about local optimization!

[16] As an exercise, show that you cannot do better by dividing T into many time intervals instead of just two.

[17] The reason has to do with the almighty dollar. For physics departments in the United States to balance their budgets, introductory physics courses have to cater to engineering majors who vastly outnumber physics majors and so textbooks are written accordingly. The emphasis is on the useful and the practical. It would be interesting to find out if in other countries physics education is also skewed by financial considerations.

[18] As Feynman's high school teacher, Mr. Bader has become an almost mythic figure for generations of physics students. Probably every high school physics teacher dreams of having a student like Feynman. See, for example, https://www.quora.com/What-do-we-know-about-Mr-Bader-Feynmans-high-school-physics-teacher-who-he-mentions-in-his-lectures-and-autobiography.

[19] R. P. Feynman et al., *The Feynman Lectures on Physics*, chapter 19 in vol. 2.

[20] See *GNut*, pages 207–213.

[21] See, for example, J. M. Figueroa-O'Farrill, "Electromagnetic Duality for Children," http: www.maths.edu.ac.uk/jmf/Teaching/Lectures/EDC.html.

[22] See, for example, *QFT Nut*, chapter VI.3.

[23] Incidentally, many members of this school also belong to the "Shut up and calculate" school mentioned in chapter 3.

[24] The speed of light in air is slightly less than that in empty space, but the difference is usually negligible.

[25] The Hamiltonian is obtained from the Lagrangian by a standard maneuver in mathematics called the Legendre transform, Legendre meaning "the son in law" in French.

[26] Try as I may, I cannot bring up a mental image of this guy.

[27] After the Hamiltonian there was yet another formulation of classical mechanics, the Hamilton-Jacobi formulation. More yawns! Who cares? Indeed, except for some, notably physicists working on optics, the Hamilton-Jacobi formulation was relegated to the back burner.

[28] I don't. See *Fly by Night Physics*, Princeton University Press, 2020.

[29] Dirac was inspired by some suggestive resemblance of the classical action to certain quantum quantities.

[30] Incidentally, if you guessed the length of the path in spacetime, you're not wrong. That is exactly the action assigned to a path in Einstein's theory of special relativity. But do notice that what I am describing here also applies to the nonrelativistic world.

[31] Usually taken to be the x-axis.

[32] For readers familiar with complex numbers, the precise relation is $\mathcal{A} = e^{iS(\text{path})/\hbar}$.

[33] For readers who know some quantum mechanics, an easy introduction may be found in *QFT Nut*, pages 7ff. At a level a bit higher than the typical popular book, see *QFT ASAP*. Actually, in spite of the fact that the path integral formulation is never ever, to the best of my knowledge, taught to undergraduates, Feynman and a student at Caltech named Hibbs, who attended Feynman's lectures on this approach, had published a textbook. R. P. Feynman and A. R. Hibbs, *Quantum Mechanics and Path Integrals*. According to a rather nasty review circulating in the theoretical physics community, this book contains "Not a thought of Hibbs, and not a word of Feynman." Be that as it may, the book probably holds the record for the number of typographical and other errors in a physics textbook. I do not doubt that Feynman never even glanced at the manuscript. Forty-five years after the original publication of the book, and after the death of both authors, D. F. Styer performed an immeasurable service to humanity by publishing an "Emended Edition." According to the Amazon website, "This emended edition of the original 1965 publication corrects hundreds of typographical errors and recasts many equations for clearer comprehension. It retains the original's verve and spirit, and it is approved and endorsed by the Feynman family."

[34] For a semipopular account, see *QFT ASAP*.

[35] Recall that the energy unit erg is defined to be 1 gram (centimeter per second)^2, that is, twice the kinetic energy of a mass of 1 gram moving with a speed of 1 centimeter per second. In other words, $\hbar \simeq 10^{-27}$ erg second.

[36] Indeed, we already mentioned this in chapter 5.

[37] Namely, the extremum is reached when the derivative of the function vanishes, that is, when change equals zero.

[38] As you could readily imagine, that the action is sometimes maximized and sometimes minimized was particularly troublesome during its controversial history. In the Académie Française, the Chevalier D'Arcy rightfully demanded of Pierre de Maupertuis to explain how the Divine Entity could be the "Great Squanderer" as well as the "Great Saver."

[39] See *QFT ASAP*.

[40] The pioneers of string theory include Y. Nambu, T. Goto, and A. Polyakov. Try Googling the Nambu-Goto action.

[41]Of course, constructing the action requires learning some "machinery." Historically, the discovery of the needed machinery, namely the gauge potential mentioned in chapter 7, was not without struggle.

[42]Intriguingly, Attilio Palatini showed that the action for Einstein gravity could also be recast into what is known as the Palatini formulation, which some would say is more revealing than that given by Hilbert and by Einstein. See *GNut*, page 395.

[43]I am obliged to gloss over the history of string theory, which grew out of something known as the dual resonance model.

[44]See *GNut*, page 396, for Einstein's denunciation of "nostrification" and the "wretchedness of humanity."

[45]For instance, viscosity in fluid flow is due to momentum transfer between a fast moving layer and a slower moving layer. Similarly, if we do not include the energy and momentum carried away by electromagnetic waves emitted by a charged particle, the resulting equation of motion would include a friction term.

[46]Recall the example of the Gaussian integral mentioned in chapter 8.

[47]For an inkling of how this could possibly work, see pages 739–741 in *GNut* and chapters N.2, N.3, and N.4 in *QFT Nut*.

PARTING COMMENTS

A tapestry of ideas woven with many threads

Physics is a tapestry of ideas, woven with many threads by many hands. Depending on what you think would qualify as an idea, the actual number may be in the hundreds of thousands, or millions. Evidently, some of these ideas stand out as permanent and profound, others less fundamental.

Ideas exist on many levels. In discussing the conservation of energy, we could detail how one form could be transformed into another form. Fascinating stuff! Take hydroelectric power. From the photons generated by nuclear reactions inside the sun lifting the water molecules up against the earth's gravity, through the subsequent conversion of gravitational potential energy into the kinetic energy that drives the turbine. So on and so forth, till eventually the electric current in our kettle heats the water molecules and causes a phase transition from liquid to gas, which drives a whistle that leads to a tiny electric current in our brains.

Each of these steps involves nontrivial and substantial physics that took decades and centuries to unravel. Yet, to say that the conservation of energy is due to the invariance of the laws of physics under translation in time clearly takes us to another level, as Einstein rhapsodized.

A priori, how could we have even dreamed that the myriads of processes we see around us could be reduced to four interactions, and then for sure to three and possibly to two? (And hopefully to one!) How could we have hallucinated that the three nongravitational interactions all depend on transformations in an internal space that vary from place to place in spacetime? And that this variation in spacetime necessitates the introduction of gauge bosons superintended by a branch of mathematics, group theory, that its inventors in the 19th century said that physicists could never use?

Friends, that is the wonder and enjoyment of theoretical physics! Not some of the ephemera, the latest and the hottest, which bombard us every day. I recognize that this book may have a wide range of readers whose interests and outlook may not coincide with my own, but I would think that most of these ten ideas outlined here, give or take, would undergird, explicitly or implicitly, any serious discussion of physics. And any such discussion would have to presuppose that the (physical) world is comprehensible.

I do not mean to say, of course, that in the last few decades earthshaking discoveries had not shaken physics. To the contrary, some of these may turn out to be be extremely important, perhaps leading both to a deeper understanding of physics and to technological applications (just like the discovery of electron spin discussed in chapter 3). To entertain only one example, the discovery that the universe is expanding faster and faster (as was mentioned in chapter 6), contrary to what physicists once almost universally believed. Well, I choose to distinguish between important and foundational in this book. But who knows, the accelerating expansion and the dark energy that drives it may somehow be linked to one of the outstanding puzzles in particle physics, the triplication of matter in the universe in spite of the fact that one generation of quarks and leptons would suffice (as was mentioned in chapter 7). A linkage between the cosmological expansion and the family problem would be mind blowing indeed.*

I offer you here a few parting comments, rather than concluding remarks. This book does not have a conclusion as such, but perhaps it would be enjoyable to run over once again the ten foundational ideas that I regard made physics what it is.

1. The physical world is comprehensible.
2. The laws of physics are the same here, there, and everywhere, the same yesterday, today, and tomorrow.
3. The world is quantum.
4. Quantum fields forever.
5. The deeper we go, the more symmetric the laws look.
6. Einstein, the exterminator of relativity and the choreographer of spacetime.

---

*The number 3 seems to pop up in physics a lot. To do physics, we need three fundamental units to measure mass, length, and time respectively, and three fundamental constants ride to our rescue: Newton's G, Planck's $\hbar$, and Einstein's c. (For example, once we have a unit for time, then c obviates the need for a unit for length.) We live in 3-dimensional space, there are three colors in the world of quarks, and the matter in the universe (the dirt so to speak) is triplicated.

7. Unity of forces in the universe.

8. The Creator speaks the language of mathematics.

9. Entropy and thermal agitation.

10. Physics is where the action is.

Many of these ideas are forcibly intertwined, as I emphasized throughout. Among these, entropy stands out for intertwining the microscopic world with the macroscopic world, a linkage not achieved without a titanic struggle spanning centuries and ending in human tragedy. Later, the surprising connection between entropy and information is opening new windows, in particular the promise of quantum computation, which itself is twisted with the entanglement characteristic of the quantum world (as discussed in chapter 3).

Echoing back to the sentiments of Newton, Kelvin, and Rosenfeld expressed in chapter 1, I see the theoretical physics community populated by optimists and pessimists, as is always the human condition. The optimists feel that the final theory is at hand, and fervently hope that it will arrive before they in particular pass from this vale of tears. The pessimists think that our dream of a final theory remains elusive. Perhaps it will forever remain a dream, with layers upon layers as in an infinite series. But even that may came to an abrupt halt when we are contacted, and if instructed instead of enslaved, by superior beings.

Make your own list

I would like to reiterate what I said in the prologue. I do not doubt for a second that other physicists would come up with other lists, but I would be much surprised if they don't overlap with my list significantly. As I said, people who don't like my list are cordially invited to make their own lists and write their own books.

Yes, lots of Nobel prizes are of only transient or local interest. Most people are surprised when I tell them that one of the early physics prizes was given for a switch that turns the light at lighthouses on and off automatically. No doubt this was a valuable and important idea, but that is not what I want to talk about in this book. No one could plausibly argue that quantum mechanics and special relativity sit on the same level as an automatic switch for lighthouses. Indeed, GPS, which relies on both quantum mechanics and special relativity, made lighthouses[1] increasingly obsolete. Again, the distinction between important and foundational.

One criterion for what I regard as an idea is something that occurred to very few people, perhaps to only one person. For example, that the space-time around massive objects is curved and even warped. Now, that's an idea! Another pervasive criterion, as in some of my other books, is communication with extraterrestrials. Anything that depends on our having ten fingers, or markings on a tube of mercury, cannot possibly have anything to do with the fundamental laws of the universe.

I have stayed away from speculative ideas that are not yet established empirically, notably string theory. I have also omitted mention of speculations on the nature of the dark energy and dark matter. Nevertheless, I have discussed briefly grand unified theory as part of the drive toward unification, which is certainly an empirically established reality in physics. Also, the seamless group theoretic fit of the theory has convinced me and many others that there is more than a grain of truth in grand unified theory, even though it suffers from some tension with observations. Incidentally, grand unified theory has been around for almost fifty years, as has string theory, depending on when and what you consider as its germination.

The seven pillars of wisdom

Einstein once listed the seven principles that guided his physics: unity, simplicity, generalization, logical parsimony, deterministic causality, completeness, and the continuum. This book touches all seven.

Physics at the fundamental level, at least the part we understand and have established empirically, has turned out to be truly and wonderfully simple. The fundamental laws are as simple as possible, but not any simpler, as Einstein had wished. Needless to say, it is simple only if you understand the language it is expressed in. And of course we cannot say whether the part we don't understand, notably the possible unification of gravity with the other three interactions, is simple or not until we understand it. Only time could tell. The weak interaction ("God's mistake" as per Nambu) surely did not look simple during the first half of the 20th century.

That physics is surprisingly simple is responsible for evoking the wonderful feeling, shared by many physicists, that physics in graduate school turned out to be simpler than the physics they learned in undergraduate courses. This revelation of underlying simplicity was at once eye-opening and exhilarating.

The drive toward simplicity is closely linked to the drive toward unity, and to the natural desire to generalize. We saw spacetime physics generalized to Einstein gravity with its curved and dynamical spacetime. And then we saw the abelian gauge theory of electromagnetism generalized and promoted to the nonabelian gauge theory of the electroweak, the strong, and possibly the grand unified interactions. In spite of the nattering nabobs of negativism, even if grand unified theory in its present form is not in total agreement with observations, the drive toward grand unification, and ultimately unification with gravity, is surely in the right direction and in accord with what Einstein had wanted. Do some negativists prefer to see the fundamental structure of the universe constructed in several disjoint pieces?

By logical parsimony, Einstein probably meant, I would surmise, not only the economy of symbols, but the economy in the requisite logical steps. What could be more parsimonious than an action that could be written down on a piece of napkin?

Of the items on Einstein's list, completeness is something that could always be negated tomorrow, but the fact remains that for well over a century, we have not seen anything proven to sit outside the existing framework, or at least, cannot be accommodated readily. I may have spoken too early here about the dark side of the universe, the most likely arena of unexpected surprises. Yet up to now at least, the mainstream ideas about dark energy and dark matter could all fit within the existing framework of physics.

Then we turn to the most problematical items on Einstein's list: deterministic causality. What did he mean by mashing two distinct concepts together? Causality is built into quantum field theory, and thus is guaranteed in any process it could describe. Cause precedes effect. In quantum physics, the evolution of the wave function from one instant in time to the next is fully deterministic, but the notorious "collapse of the wave function" when it is measured is probabilistic, and the wave function itself encodes the probability amplitude.

If by "deterministic" Einstein meant the determinism of classical physics, then deterministic physics is, alas, no more, as was convincingly shown by the experimental violation of Bell's inequality discussed in chapter 3, which Einstein, sadly, did not live to see. Causality, however, lives on.

Finally, the continuum, as in the concept of field "totally loved," as Oppenheimer put it, by Einstein, and I might add, his intellectual descendants! All praise to the interweaving dance of quantum fields, as was sung in chapter 4.

There you have it: a physics built on unified and causal quantum fields, utterly simple, I am tempted to say even frighteningly simple, built on logical parsimony and step-by-step generalization, apparently complete, but resoundingly probabilistic. From this physics, made "as simple as possible but not any simpler," our familiar deterministic classical world, beloved though illusionary, emerges.

Which is more satisfying?

Since I write both textbooks and popular books, people often ask me which is more fun and satisfying to write, or more simply, which do I prefer to write. The glib and easy answer, but nevertheless truthful, is that whenever I am in the last stages of finishing a book, I would have preferred to be writing the other kind.

More seriously, I have thought much about this question, and have also discussed this issue with friends and colleagues. Textbooks are clearly important. Physics is difficult enough, and without constant distillation and weeding out the misleading and clarifying the confusing, would be impossibly difficult. Some of what is considered important also shifts from one era to the next. Books on quantum field theory, for example, have remained more or less the same thickness since I was a student. On the other hand, there is also the competing view that textbooks cater to those already committed to going into physics, while popular books could move young people to go into physics, which at its most exalted level is more a calling than a profession. I agree. Over the years, I have run into a hopefully endless stream of physics students who told me that they were inspired by my *Fearful Symmetry* as undergraduates. Rob Phillips, a friend of mine and a professor at Caltech, spoke to me about this issue with particular emotion, recalling that he decided to skip undergraduate school and while he worked as an electrician read a library full of popular books on physics (as well as textbooks to get into graduate school). He was sustained by popular books over those years of freedom from school. (Incidentally, I was too, albeit for a shorter time, when my family immigrated to Brazil and I did not go to school for a year[2] during which I read a few popular books about physics.)

So, too, in this book, I tend to favor the future over the past. I hope that going over the foundation ideas of physics with you will inspire some young readers to consider this challenging but rewarding field as a lifetime pursuit.

Notes

[1] It so happened that my third son's elementary school was at the end of Lighthouse Road, which makes it probably one of the few schools in the United States with a view of the Pacific Ocean.

[2] Some of my friends feel that I was lucky not to have my nascent interest in physics possibly squashed by a dull teacher.

ACKNOWLEDGMENTS

I am grateful to Nathaniel Craig, Eric DeGiuli, Joshua Feinberg, David Gross, Jessica Howard, Greg Huber, Steve Lamoureaux, Connor Lindquist-Carrillo, Nadie LiTenn, Geoffrey Penington, Grant Remmen, Lewis Robinson, and Mark Weitzman for commenting on various pieces of the manuscript. Ingrid Gnerlich, who has worked on all my Princeton University Press books, has been enthusiastically supportive since the inception of this project. I also would like to thank my copy editor, Jennifer Harris, my proofreader, Stephanie Sakson, and my production editor, Karen Carter, for the arduous task of converting a manuscript to a book. I have benefited from the support and presence of Janice and Max. This book was mostly written in Santa Barbara, California, but bits and pieces were also written in Stockholm, Sweden; Honolulu, Hawaii; Bariloche, Argentina; Valdivia, Chile; Taipei, Taiwan; Beijing and Shanghai, China. The galley proofs were corrected in Reykjavík, Iceland. I thank my various hosts for their hospitality.

BIBLIOGRAPHY

Books by the author

I refer quite often, naturally, to the textbooks and popular books I have written, as listed here chronologically below according to the following abbreviations: *Unity, Fearful, Toy, QFT Nut, GNut, Group Nut, G, FbN,* and *QFT ASAP*. All are published by Princeton University Press, except as noted.

Unity of Forces in the Universe, World Scientific, 1982.

Fearful Symmetry: The Search for Beauty in Modern Physics, Macmillan, 1986; Princeton, 2016.

An Old Man's Toy: Gravity at Work and Play in Einstein's Universe, Macmillan, 1989; retitled as *Einstein's Universe: Gravity at Work and Play*, Oxford University Press, 2001.

Quantum Field Theory in a Nutshell, 2003, 2010.

Einstein Gravity in a Nutshell, 2013.

Group Theory in a Nutshell for Physicists, 2016.

On Gravity: A Brief Tour of a Weighty Subject, 2018.

Fly by Night Physics, 2020.

Quantum Field Theory, as Simply as Possible, 2023.

Books by other authors

Some of these books were mentioned explicitly in the text; others were not. They are all recommended for further reading. A few are textbooks, but many are popular books or biographies.

R. Baierlein, *Newton to Einstein: The Trail of Light: An Excursion to the Wave-Particle Duality and the Special Theory of Relativity*, Cambridge University Press, 2001.

B. R. Brown, *Planck: Driven by Vision, Broken by War*, Oxford University Press, 2015.

J. Bjorken and S. Drell, *Relativistic Quantum Mechanics*, McGraw-Hill, 1964.

J. Bjorken and S. Drell, *Relativistic Quantum Fields*, McGraw-Hill, 1965.

I. Duck and E.C.G. Sudarshan, *Pauli and the Spin-Statistics Theorem*, World Scientific Publishing 1998.

G. Farmelo, *The Strangest Man: The Hidden Life of Paul Dirac, Mystic of the Atom*, Basic, 2011.

R. P. Feynman and A. R. Hibbs, *Quantum Mechanics and Path Integrals*, McGraw-Hill, 1965.

S. Gasiorowicz, *Elementary Particle Physics*, Wiley, 1966.

R. Golub and S. Lamoreaux, *The Historical and Physical Foundations of Quantum Mechanics*, Oxford University Press, 2023.

M. Le Bellac, *The Quantum World*, World Scientific, 2013.

R. B. Leighton, *Principles of Modern Physics*, McGraw-Hill, 1959.

D. McIntyre, C. Manogue, and J. Tate, *Quantum Mechanics: A Paradigms Approach*, Pearson, 2012.

A. Pais, *Subtle Is the Lord*, Oxford, 1982.

J. J. Sakurai, *Invariance Principles and Elementary Particles*, Princeton University Press, 1964, 2016.

J. J. Sakurai and San Fu Tuan, *Modern Quantum Mechanics*, Addison-Wesley, 1994.

J. J. Sakurai and Jim Napolitano, *Modern Quantum Mechanics*, Cambridge University Press, 2017.

A. D. Stone, *Einstein and the Quantum: The Quest of the Valiant Swabian*, Princeton University Press, 2015.

S. Weinberg, *The Discovery of Subatomic Particles*, W. H. Freeman, 1983.

TIMELINE

The ages at death are rounded off; I do not keep track of the month of death.

Galileo di Vincenzo Bonaiuti de' Galilei (1564–1642), 78
Pierre de Fermat (1607–1665), 58
Giovanni Domenico Cassini (1625–1712), 87
Christiaan Huygens (1629–1695), 66
Isaac Newton (1642–1726/27), 85
Ole Christensen Rømer (1644–1710), 66
Leonhard Euler (1707–1783), 76
Joseph-Louis Lagrange (1736–1813), 77
Marquis Pierre-Simon de Laplace (1749–1827), 78
Carl Friedrich Gauss (1777–1855), 78
Michael Faraday (1791–1867), 76
Niels Henrik Abel (1802–1829), 27
John Waterston (1811–1883), 72
Rudolf Julius Emanuel Clausius (1822–1888), 66
Bernhard Riemann (1826–1866), 40
James Clerk Maxwell (1831–1879), 48
Josiah Willard Gibbs (1839–1903), 64
Ludwig Eduard Boltzmann (1844–1906), 62
Oliver Heaviside (1850–1925), 75
Hendrik Antoon Lorentz (1853–1928), 75
Max Karl Ernst Ludwig Planck (1858–1947), 89
Hermann Minkowski (1864–1909), 45
Karl Schwarzschild (1873–1916), 43
Albert Einstein (1879–1955), 76
Paul Ehrenfest (1880–1933), 53
Otto Sackur (1880–1914), 34
Max Born (1882–1970), 88
Amalie Emmy Noether (1882–1935), 53
Niels Henrik David Bohr (1885–1962), 77
Otto Stern (1888–1969), 81
Walther Gerlach (1889–1979), 90
Hugo Martin Tetrode (1895–1931), 36
Wolfgang Ernst Pauli (1900–1958), 58
George Eugene Uhlenbeck (1900–1988), 88
Samuel Abraham Goudsmit (1902–1978), 76
Eugene Wigner (1902–1995), 93
Julius Robert Oppenheimer (1904–1967), 63
Ralph Kronig (1904–1995), 91

Hendrik Brugt Gerhard Casimir (1909–2000), 91
Chien-Shiung Wu (1912–1997), 85
Claude Elwood Shannon (1916–2001), 85
Richard Phillips Feynman (1918–1988), 70
Julian Seymour Schwinger (1918–1994), 76
Murph Goldberger (1922–2014), 92
Philip Warren Anderson (1923–2020), 97
Abdus Salam (1926–1996), 70
Tsung-Dao Lee (1926–2024), 98
Robert Laurence Mills (1927–1999), 72
Murray Gell-Mann (1929–2019), 90
Steven Weinberg (1933–2021), 88
James Daniel "bj" Bjorken (1934–2024), 90
Sidney Coleman (1937–2007), 70
******

INDEX

Page numbers in *italics* indicate figures, footnotes, and tables

313